AN INTELLIGENT SYSTEM FOR
TRIBOLOGICAL DESIGN IN ENGINES

Aims & Scope

The Tribology Book Series is well established as a major and seminal archival source for definitive books on the subject of classical tribology. The scope of the Series has been widened to include other facets of the now-recognised and expanding topic of Interface Engineering.

The expanded content will now include:
• colloid and multiphase systems; • rheology; • colloids; • tribology and erosion; • processing systems; • machining; • interfaces and adhesion; as well as the classical tribology content which will continue to include • friction; • contact damage; • lubrication; and • wear at all length scales.

TRIBOLOGY AND INTERFACE ENGINEERING SERIES, 46
EDITOR: B.J. BRISCOE

AN INTELLIGENT SYSTEM FOR TRIBOLOGICAL DESIGN IN ENGINES

XIANGJUN ZHANG
State Key Laboratory of Tribology
Tsinghua University, Beijing
People's Republic of China

CHANGLIN GUI
School of Mechanical and Automotive Engineering
Hefei University of Technology, Hefei
People's Republic of China

2004

ELSEVIER

Amsterdam – Boston – Heidelberg – London – New York – Oxford
Paris – San Diego – San Francisco – Singapore – Sydney – Tokyo

ELSEVIER B.V.
Sara Burgerhartstraat 25
P.O. Box 211, 1000 AE
Amsterdam, The Netherlands

ELSEVIER Inc.
525 B Street, Suite 1900
San Diego, CA 92101-4495
USA

ELSEVIER Ltd
The Boulevard, Langford Lane
Kidlington, Oxford OX5 1GB
UK

ELSEVIER Ltd
84 Theobalds Road
London WC1X 8RR
UK

First edition 2004

Library of Congress Cataloging in Publication Data
A catalog record is available from the Library of Congress.

British Library Cataloguing in Publication Data
A catalogue record is available from the British Library.

ISBN: 0-444-51756-1 (Vol. 46)
ISSN: 1572-3364 (Series)

♾ The paper used in this publication meets the requirements of ANSI/NISO Z39.48-1992 (Permanence of Paper).
Printed in The Netherlands.

Foreword I

It is my honour to write the foreword for this book on "An Intelligent System for Engine Tribological Design".

Tribology research in China faced a rigorous challenge when it stagnated in the 1980s, as its major interests were focused on elastohydrodynamic theory. However, since the beginning of the 1990s, a new period in tribology has emerged, which is indicated by two significant trends. One is the combination of tribology with information technology, materials science, nanotechnology, and the life sciences. These intersection have given birth to micro/nanotribology, biological tribology, and control tribology, and have brought about a boom in activity and developments to this traditional discipline. The other trend is towards a closer combination with engineering science and engineering technology, e.g., modern design methodology, dynamics, and transmission technology. This has given birth to engineering tribology and tribological design, and has brought tribology out of books and laboratories, to play a more significant role in mechanical engineering and practice. This book describes one of the representative developments of this new period of tribology research in China.

Under the sustained support of the National Natural Science Foundation of China, Professor C. L. Gui and his assistants have persevered their research on engine tribological design for many years. They have combined engine tribology with information sciences, artificial intelligence technology, non-numerical algorithms, modern design technology, and dynamics, and have proposed a new methodology and technology for engine tribological design, that is, intelligent design. The intelligent system for engine tribological design comprises the main content of this book, which is a concentration of the research, knowledge, and the experiences of the authors. It is also one of the last fruits of tribology research in China. Its publication will definitely present a valuable reference book for researchers and engineers in the field of tribology. This book will also be significant in promoting further combination of tribology with engineering, and further development of tribology.

Yuanzhong   LEI

Professor, Director of the Division of Mechanical Engineering
National Natural Science Foundation of China

Foreword II

The internal combustion engine has been widely used in engineering practice. With its development increasing demands have been put on the tribological problems of the engine. This consequently prompts the research of tribological theories and practice in the engine field.

The authors creatively combined the intelligent design technology with the traditional tribological design of an engine, to present an effective approach for tribological design of an engine and also a new pattern for tribology research.

This book gives a concentration of the authors' research work on the intelligent system for tribological design in engines and collects the latest relative developments in the world. It includes 6 chapters, containing the basic contents of tribological design of an engine and its intelligent design system. It was especially written for engine design engineers and tribological design researchers, and will be helpful to their practice and research.

Finally, I also would like to express my personal sincere respects to the authors for their dedication to the technological readers abroad in the field of tribological design of engine.

Shizhu WEN

Professor, State Key Laboratory of Tribology
Tsinghua University, Beijing, China

Acknowledgements

The authors are grateful to the National Natural Science Foundation Commission of China, and Technical Development Foundation Commission, Ministry of Machinery Industry of China for their financial support.

The authors also thank their colleagues and friends for their helpful discussion and sincere support.

Contents

Chapter 4.   The Artificial Neural Network Model and its Application in ICADEDT

Chapter 5.   Gene Model-based Comprehensive Decisions for Tribological Design

Chapter 6   A Hierarchical Cooperative Evolutionary Design for Engines

Chapter 1: Engine Tribological Design Methodology and Intelligent System

1.1 A brief introduction to tribological design

1.1.1 History of tribological design and its development

In the relatively long history of civilization development for human beings, the original aim of creating and developing a machine was to realize a certain action. Thus, the essential consideration of machine design was the desired action, though the strength and friction problems should be involved. However, with the invention of the internal combustion engine and electrical motor, the velocity and transmitted power of a machine are extremely improved, consequently, the strength problems, including static strength, dynamic strength, and stiffness, are becoming more and more important. In addition to the kinematic design, the strength and stiffness demands, which generally determine the structure and dimension of a mechanical component, must be achieved for a machine. Strength design has then turned to be the heart of mechanical design. Meanwhile, strength design theories and methods have been significantly developed, and have been familiarized and mastered by mechanical engineers abroad.

Since the 60s of last century, there has been an increasing demand on the efforts to deal with the friction, lubrication, and wear problems of mechanical equipment. Tribology, emerged as an independent discipline under such an increasing demand, proposes many solutions to the involved friction, wear and lubrication problems. Here are seven principle approaches:
(1) design
(2) materials

(3) surface treatment
(4) lubricant
(5) lubricating system
(6) sealing and filtering system
(7) equipment administration and monitoring, component maintenance

Among them, only design is the most effective approach to the tribology problems. It is because that,
(1) friction and wear performances of a machine are determined by its design;
(2) authority of the design blueprint and the relevant technology employed in the design are widely accepted during production.
(3) the considerations in design are comprehensive and systematic.

Dr. H. Peter Jost is the first scholar who defined the new item, Tribology, as the discipline involving friction, wear and lubrication, and promoted its development all over the world. He is also the first man to propose the concept of tribological design. In the preface of *Principles of Tribology* edited by prof. J. Halling in 1975, Dr. Jost wrote that, "The main reason behind this development, which is also the principle reason for the importance of this book, is the recognition of the close interrelationship between tribological design principles and practices on the one hand, and their economic effect on the other. The days of single disciplinary designs and of design by trial and error are gone forever. Modern products must, during their design stages, have incorporated all the factors that lead to a satisfactory control of friction and prevention of wear."[1]

We completely approve of Dr. Jost's point of view about the signification and valuable role of tribological design. So, in the recent two decades we have been devoting ourselves to the research on the tribological design theories and methods of the internal combustion engine, and we will go on with our research in the future. Our main fruits in research work are presented in this book now.

In the summer of 1985, ASME and USA energy and commercial bureau held a special issue discussion about the establishment of the tribological database. The principle of the database system aims not only to simply restore information, but also to propose instructions to engineers on how to apply tribology knowledge to machine design, operation and administration. The proposed system was planned to adopt artificial intelligent technology, store topics and knowledge in the form of the "IF...THEN..." rule, and perform the mission through a "dialog and negotiation " mechanism between the user and the terminal. As a reader of the system menu and literature, the mechanical engineer submitted a design mission to the system, then the system turned into a consulting work mode through a "introduction module", which checked, compared and synthesised the former solved similar problems, and made design decision for user to deal with the current design problem. The system was able to simulate domain expert to perform the tribological design of a component basing on the combination of the tribological knowledge and database technology. Therefore, the proposed system was expected to be a consulting computer expert system for tribological design.

Three years later in 1988, in the 15th Leeds-Lyon Symposium on Tribology, tribological design had become a main subject, and researchers from all over the world presented and communicated their work.

In China, under the proposal of prof. Xie Youbo, a member of Chinese Academy of Engineering, a symposium on tribological design was held in Nanjin in April 1989, supported by the tribology division and mechanical design division of Chinese Mechanical Engineering Institute. The subjects addressed were the signification, definition, object and method of tribological design. In Oct. 1991, the 1st Chinese Triblogical Design Conference was held in Shenyang, the submitted papers were collected and published publicly. Meanwhile, China National Science Foundation Committee (NSFC) added a new item titled "tribological design and its database" within the entry "mechanics discipline", so that scholars who are interested in this domain might be able to obtain the financial support. Three years later in Oct. 1994, the 2nd Chinese Tribological Design Conference was held in Beijing.

It can be summarized that, all the papers presented to the two conferences mentioned above and those presented to the 15th Leeds-Lyon Symposium have the two common features:

(1) Intelligent computer aided design method was suggested as a tribological design method to combine the analysis calculation and the domain expert knowledge based on database technology.

(2) The involved research object was limited in the level of design of mechanical component, however, tribological design of an entire machine has not been reported.

1.1.2 Inherent nature of tribological design

Up to now, an authoritative definition of tribological design has not been found in the literature on tribology. In our opinion, presenting a definition of tribological design is not of upmost importance. The inherent nature of tribological design should be well defined, and should essentially include the following 5 aspects, i.e. aims, theory, method, criteria and database.

Aims – specific engineering requirments

The aims of tribological design of mechanical equipment derive from,

(1) machine breakdown or the unbearable maintenance cost due to the low wear life of components, i.e., the aim to prolong wear life.

(2) demands to decrease frictional loss or energy loss, and enlarge effective power output , i.e., the aim to decrease frictional power loss.

(3) scratch and abrasion of friction pair which badly threaten the normal operation of machine, i.e., the aim to prevent scratch and abrasion.

(4) loss of accuracy due to the wear of components, i.e., the aim to perform anti-wear design for accuracy.

(5) reliability problem caused by the friction and wear of components, i.e., the aim to improve the reliability of machine.

(6) impermissible vibration and noise generated by friction and wear of components, i.e., the aim to satisfy the vibration and noise demands.

(7) movement dislocation and hysteresis due to friction and wear of components, i.e., the aim to keep movement accurate and on time.

(8) demands of special lubricating and frictional materials for equipment in special environment, i.e., the aim to meet special environmental demands.

(9) seal problems due to component wear, i.e., the aim of qualified seal.

(10) impermissible thermal deformation of components attributed to frictional heat, i.e., the aim to control thermal deformation.

It can be seen from the above that, all the tribological design aims are related to certain equipment or frictional pair in a concrete environment or work condition. Therefore, tribological design is a design process aiming at a concrete object, i.e., there does not exist a tribological design without a concrete object. Besides, the objects of tribological design fall into two categories, one is the general frictional pairs including wheel gear, worm gear, chain gear, bush bearing, rolling bearing, guide way, knife tool, and die, etc., the other is the machine equipment including internal combustion engine, electric generator, machine tool, rolling machine, etc.

Theory — based on the current tribology principles

The current tribology principles mainly include,
(1) lubrication mechanics;
(2) lubricant, tribo-chemistry and boundary lubrication principles;
(3) physical phenomena, mechanism and function that involved in friction, i.e., tribo-physics

(4) surface structure and properties of metal materials in friction, i.e., tribo-metallurgy

(5) contact mechanics in friction, i.e., tribo-mechanics

(6) friction mechanism

(7) wear principles

Method — establishing the analytic or experimental relationship between aims and influence factors

A typical unit tribological system can be described as Fig.1-1, which is composed of three basic elements, i.e., structure, materials and lubrication.

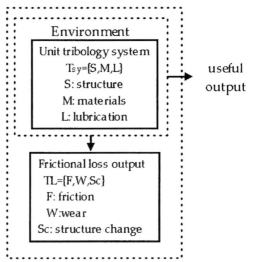

Fig.1-1 A typical unit tribological system

Under practical circumstances, two other elements are also involved. One is the input working parameters, the other is the working condition. With the five elements, the unit tribological system produces useful output as well as frictional power loss. Among the five basic elements, working condition and working parameters are basically invariant from the viewpoint of design, while the structure, materials and lubrication elements are variable. Therefore, its tribological design is conducted to obtain maximum effective output at the price of minimum frictional power loss through optimization of variable elements (i.e., structure, materials and lubrication elements) within a certain working condition where the working parameters are almost constant.

The minimum frictional power loss demanded here is coincidental with the aim (2) described above. To realize all these aims in engineering practice, it is necessary to establish the analytic or experimental relationship between the aims and the structure, materials and lubrication elements involved in a tribological system based on the tribology principles, then some of the elements are modified to satisfy the design aims. Therefore, establishment of analytic expression and experimental rules to represent these tribological relationships become the primary work of a tribological design. Since the tribological design of the internal combustion engine is the subject of this book, the relevant analytic expression and experimental rules of an engine are described in detail in Chapter 2.

Criteria — establishing design criteria to meet the design demands

Similar to the engineering design, like strength design, design criteria and rules are also necessary for tribological design. For example, in the tribological design, to avoid scratch and abrasion, preventing breakage of oil film is a primary consideration. Thus, it is essential to establish a criterion to determine whether the oil film is broken or not. During design process, the corresponding design scheme will be modified continuously until the criterion is satisfied.

Database

It has been demonstrated a complete database including lubricant, anti-friction & anti-wear materials and surface coatings provides engineers a reliable foundation to perform scientific and effective tribological design.

1.2 Tribological design for an internal combustion engine

The continuing evolution of the internal combustion engine places increasing demands on its durability, reliability, fuel efficiency, and exhaust emissions, in which tribology plays an increasingly important role. With the aim of increasing technical and economic performance[3-5], the tribological design of an engine includes the following aspects:

(1) Avoidance of severe wear in friction pairs, to increase the reliability of an engine.

(2) Prolonging the wear-resistant life of friction pairs, to increase the durability of an engine.

(3) Reduction in the frictional power loss of an engine, to improve its mechanical efficiency and effective power output.

(4) Reduction in the fuel and oil consumption, to meet the demands of economical operation and environmental protection.

It can be seen from the above that wear, frictional power loss, and fuel and oil consumption are the major problems encountered in the tribological design of engines. These problems manifest themselves mainly in three main friction pair groups, which include the 10 friction pairs shown in Fig.1-2[4].

In general, the piston assembly, bearings, and the valve train are the most critical engine components for a tribological investigation, and the tribological performances of the three main friction pair groups play a significant role in an engine's performance. For example, their frictional power loss is greater than 75% of the total mechanical loss, as shown in Fig.1-3[3,5].

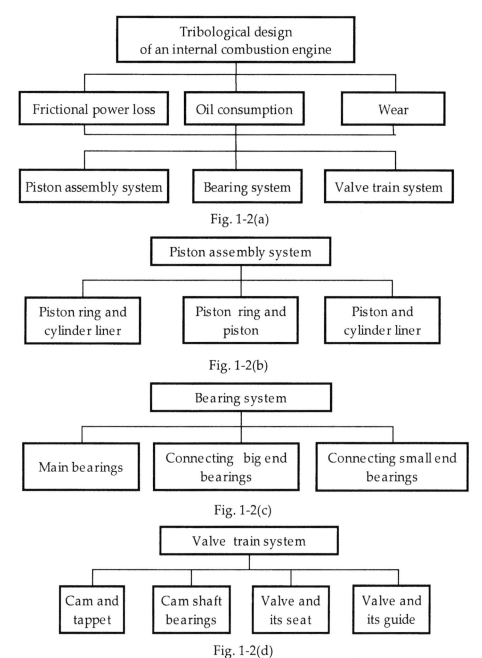

Fig. 1-2(a)

Fig. 1-2(b)

Fig. 1-2(c)

Fig. 1-2(d)

Fig. 1-2. The 10 friction pairs in the tribological design of an internal combustion engine, which can be divided into three main friction pair groups, i.e., the piston assembly system, the bearing system, and the valve train system. The wear, frictional power loss, and oil consumption are the major problems encountered.

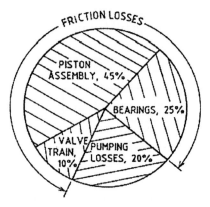

Fig.1-3. Distribution of mechanical losses in an internal combustion engine. The frictional power loss can be greater than 75% of the total mechanical loss, of which, the piston assembly accounts for more than half of the total engine friction.

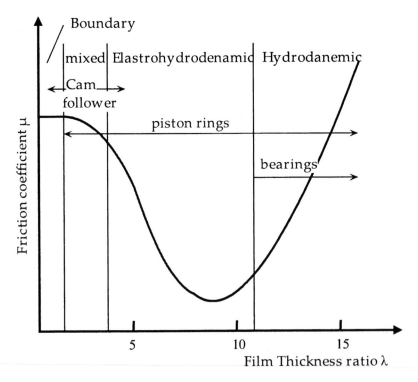

Fig. 1-4. The Stribeck diagram identifying the lubrication regimes of engine components, including the bearings, piston assembly, cam and follower.

The three engine components mentioned above present sophisticated tribology lubrication modes, covering the range from boundary lubrication to rigid hydrodynamic lubrication, as shown in the Stribeck diagram of Fig.1-4[3]. Therefore, their tribological performances have been the focus of much attention over the past few decades, with good progress being achieved[5-7] on the extensive tribological design analysis required for internal combustion engines.

From a tribological point of view, three of the aspects an engine designer needs to consider, i.e., structural design, lubrication design, and material design.

The structural design component comprises three parts:
(1) Structural design of the friction pairs, *e.g.*, the structural form, dimensional parameters, and surface topography of each friction pair.
(2) Structural design of the auxiliary systems, *e.g.*, the lubricant supply system, the filtering system, and the cooling system, and
(3) Structural design of the entire set of internal combustion engine parameters, *e.g.*, the stroke-bore ratio and normal rotational speed.

The lubrication design component also includes three parts, which are:
(1) Selection of the lubricant and additives.
(2) Design of the lubricating and filtering system and the cooling system.
(3) Design of the monitoring and control system of the supply condition and of the change in quality of the lubricant.

The material design component can be classified into two parts.
(1) Selection of the surface materials of the friction pairs, and
(2) Design of the surface coating materials of the friction pairs.

To implement these design components, a major ingredient in the design process is the use of various design analysis methods, including the use of mathematical modelling to calculate tribological performances of the components. However, for a practical design problem, these detailed design analysis methods alone are insufficient, as they are only the primary stages of the design process. A typical design process for a mechanical component[8] is illustrated in Figure 1-5.

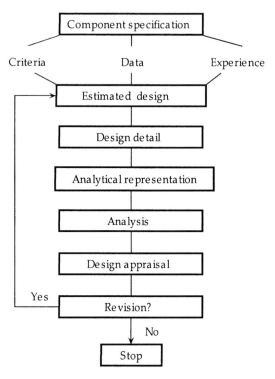

Fig. 1-5. The design process of a mechanical component.

For a design process, the first step is the "component specification" stage, and this involves establishing the design criteria, the data on the material properties, defining design constraints, and checking the required dimensions, in addition to the application of the intelligence and experience of the human designer. The result is an estimated (or initial) design. The next step is to carry out some form of analysis, the results of this analysis are interpreted, and if necessary, the design is revised. The revised design is analysed once again, and in this way, the design process continues until predicted measures of behaviour (e.g., oil film thickness, stress, and temperature) reach satisfactory, allowable values.

Furthermore, three stages can be identified in the design process:
(1) The development and refinement of mathematical models.
(2) The incorporation of these models into a methodology that can be used by designers, and
(3) A change in the focus of the design process from components to the system.

It is obvious for engine tribological design that we still have a long way to go from the design analysis stages to the design methodology stage, and from the component design stage to the system design stage.

However, we have to ask, "What will the engine tribological design methodology be?"

1.3 An investigation into tribological design methodology for engine

As the tribological analysis of engines is not able to provide an insight into the design problem solving process, in the following section, a piston assembly system will be analysed first, to provide an example of how to establish what types of design methods are being used, and should be used, in the tribological design of an engine. For this purpose, all the contributing factors to the problem, and their corresponding design strategies, are listed in Table 1-1.

It can be seen from Table 1-1 that the tribological design of a component is part analytical and part empirical. The designer works using design strategies. Most of these do not directly derive from tribological theoretical analyses or design standards, but are built up through the domain experts' many years of accumulated experience.

Table 1-1. The tribological factors influencing a piston assembly system and their design methods.

First grade influencing factors	Second grade influencing factors	Third grade influencing factors	Corresponding design strategies	Classification of the design methods
Lubricating state	Environment -al factors	Temperature field	a. Modify structure of chamber (Fig.1-6 and Fig.1-7) b. Cut a heat-insulation slot (Fig.1-8) c. Special cooling design for piston(Fig.1-9)	Domain experts' advice and theoretical analyses
	Working condition factors	Pressure of cylinder liner and moving speed of piston	Slightly modify engine normal rotational speed and working pressure under constant working power. (Technical support from domain expert is needed.)	Domain experts' advice and theoretical analyses
	Structural factors	Working capacity of cylinder V_h and cylinder number i	Decrease i to increase V_h under the constant production of V_h and i	Domain experts' advice and theoretical analyses
		a. Stroke-bore ratio. b. Profile dimensions of piston and piston rings c. Width of rings d. Surface roughness. e. Elasticity of rings.	a. Optimize value of stroke-bore ratio to minimize frictional power loss. b. Increase height of the profile. c. Decreasing width of the rings. d. Decrease surface roughness. e. Decrease elasticity of rings to reduce power loss balancing with the increase of oil consumption.	

Lubricant factors	a. Viscosity b. Viscosity-temperature property c. Viscosity-pressure property	a. Select lower viscosity grade lubricant. b. Select lubricant with higher viscosity-temperature property additive	Standards of lubricating oil, domain experts' advice, test data and theoretical analyses
Material factors	a. Coating material b. Anti-frictional properties	a. Make a gas-phase deposition anti-friction film on the surface of rings. b. Make an ionization embedded anti-friction film on the surface of rings.	Domain experts' advice and test data
Structural factors	Number of piston rings	Reduce the number of gas rings.	Experts' advice and theoretical analysis.

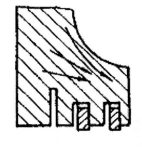

Fig. 1-9. A special cooling design for a piston.

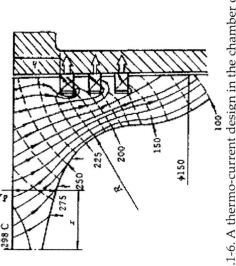

Fig.1-6. A thermo-current design in the chamber of a piston.

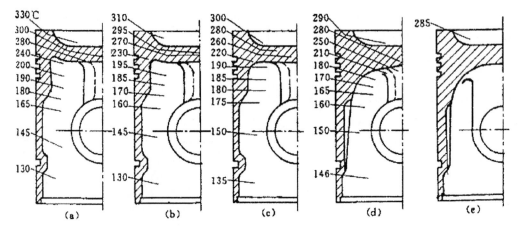

Fig.1-7. The influence of different chamber designs on the temperature field

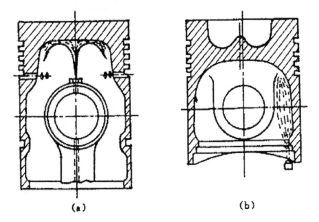

Fig. 1-8. A heat-insulation slot in a designed piston

For example, the temperature fields of the piston and cylinder liner play an important role in the engine's frictional power loss and scuffing problem, for which four practical design strategies are illustrated in Figs. 1-6–1-9. All of these obviously derive from the domain experts' empirical knowledge.

On the other hand, numerical analysis technology can overcome many of the limitations of design experience. It can analyse the concrete reasonable dimensional parameter values required for detailed design, provide quantitative information on the often conflicting design requirements by which rational compromises can be made, and even expand the design scope to analyse new materials and geometries. Taking the design strategies listed in Figs.1-6–1-9 as examples, their quantitative effects on the performance of an engine needs to be predicted using a sophisticated lubrication analysis and a Finite Element Analysis (FEA) of the temperature field, in which the material factors, structural factors, and viscosity–temperature property of the lubricant must be taken into account. The simulated quantitative effects eventually enable the engine designer to make sound design decisions that which design strategy is best for the current design.

It can be concluded therefore, that the tribological design of an internal combustion engine relies on a close combination of the domain experts' empirical knowledge and the tribological numerical analysis. The former is a qualitative approach, and the latter is a quantitative approach. They are both essential to enable an engine designer to make decisions regarding design problems. The domain experts' experience provides instructive strategies, and the numerical analysis provides exact simulations and the basis for quantitative decisions. It has also been suggested by Adam[8] that, "To be most effective, predicative techniques must be integrated into an overall product engineering design strategy". The result of this combined or integrated approach is an intelligent computer-aided design (ICAD) system.

ICAD systems have been an active field of research for almost 20 years[9-14]. Regardless of its area of application, an ICAD system is generally a knowledge-based automatic design system for complex design problems[12]. The scope of automation in intelligent design includes follows:

(1) Generating a design scheme (based on some design description, requirements, and constraints);

(2) Analysing the design scheme, by examining the properties of the design scheme generated;

(3) Evaluating the design scheme, by checking whether the design requirements and constraints are satisfied; and

(4) Redesigning, by modifying and changing the existing design scheme.

Automating the design process can be achieved by carrying out an automation search in the design space for those designs that satisfy the requirements and the constraints. Design automation or intelligent design is expected to shorten the design time and enhance the quality of the design by (a)applying highly specialized knowledge from expert sources to synthesize, or refine, the design problem, (b) allowing for the consideration of more design candidates, (c) providing more accurate information on conflicting design requirements, and (d) expanding the scope of new designs.

Therefore, an ICAD system in the field of engineering requires three essential features: various numerical analysis programs, abundant design knowledge, and most importantly, a design decision-making mechanism, which is in charge of managing, retrieving and using the various design knowledge, and invoking the analysis programs.

Numerical analysis programs are the basis of conventional computer-aided design systems. These are very application-dependent, and include different analytical tools, such as optimization techniques and FEA.

Engineering design is generally an ill-structured process, which demands a wide variety of knowledge sources, such as heuristic knowledge, qualitative knowledge, and quantitative knowledge. For a given design problem, the necessary design knowledge can also be classified as follows[13]:

(1) Generic knowledge (mathematics, physics, and common sense).
(2) Domain-specific knowledge (dimensions and tolerances in mechanical engineering).
(3) Product- or process-specific knowledge (catalogues, manuals, standards, and previous design schemes).

The design decision-making mechanism is the most important part within the ICAD architectural components, and it controls the actual execution of the design problem. The design decision-making mechanism begins with the customer requirements, and proceeds through to conceptual design, and then to the detailed design, or it begins with an initial design, and proceeds through to the evaluation analysis, to a modified design, and to the final design. Therefore, it is a multi-stage, iterative, and collaborative process, and corresponds to the various design processes.

Extensive applications of ICAD systems in engineering design have been presented in the literature. However, there have not been many applications cited for integrated machine design, and this is especially true for the tribological design of internal combustion engines.

In the field of engine tribology, much research has focused on analysing the theoretical and experimental tribological performances, including lubrication analysis, wear prediction, frictional power loss calculations, and predictions of oil consumption [2-7]. Meanwhile, abundant expert knowledge and design experience has been accumulated during the design practice. Therefore, the urgent and practical need in an intelligent system for engine tribological design is how to integrate the abundant experts' design knowledge with the existing theoretical analysis programs to carry out design making. That is, how can the intelligent system create an initial design scheme at the beginning of the process? and then how does it evaluate the design scheme; furthermore, when the design scheme's corresponding tribological performances do not satisfy the technical and economic demands of an engine, how does the intelligent system modify the design using an expert's knowledge; and, most importantly, how does it deal with the conflicts when several design schemes are proposed simultaneously, or when several design parameters interact with each other?

As the core of an intelligent system, the design decision-making mechanism not only relies on quantitative tribological numerical analysis and on the domain experts' qualitative instructions, but also needs an effective decision support mechanism. In particular, the essential problem is how to represent, obtain, and utilize the domain experts' knowledge, and then combine it with the results of the numerical analysis, and this determines the intelligence level of the ICAD system.

To implement such a decision support mechanism, a "Knowledge Model" concept is proposed in this book. Three types of knowledge models are established and practicable. Using these as well as the abundant design knowledge, various tribological numerical analysis programs and various design inference processes can be integrated into an intelligent system for tribological design of engines.

1.4 Knowledge models for ICAD system

The definition[15] of a knowledge model is that, a data model or data structure that can be adopted by an intelligent system to represent, obtain, and utilize knowledge.

The knowledge model is the basis of an intelligent system, whereby the system can represent and extract design knowledge from various knowledge sources, and integrate them with various numerical analysis programs in a seamless manner. Furthermore, this makes the design process intelligent and automatic. In this book, three types of knowledge models are adopted to implement the intelligent system for tribological engine design, i.e., symbol models, artificial neural network models, and gene models. Each of these models has its own characteristics, and each one is suitable for dealing with different types of design problem. Their applications in the tribological design will be illustrated in detail in the following chapters. A general overview of their principles and working mechanisms is presented as follows.

Symbol models

Symbol models are derived from the field of artificial intelligence, and reflect the thought mode of human beings. They represent facts, concepts, judgments, and inference procedures as a chain of symbolic language[14]. Several symbol models are widely used in artificial intelligence, including the rule model, the framework model, the semantic model, the predication logic model, and the object-oriented model, which can be referred to in the literature on Artificial Intelligence or Expert Systems[9-11].

Symbol models are traditional knowledge models used for design experience and knowledge. For example, the frictional power loss of a relative sliding friction pair is

$$N_f = PVf \tag{1-1}$$

where P is the vertical load on a sliding surface, V is the relative sliding speed, and f is the friction coefficient between the surfaces. Using Equation (1-1), the parameters influencing the frictional power loss and their functions can be determined. Furthermore, concepts can be summarized, *e.g.*, under constant P and V, to decrease the frictional power loss, the friction coefficient must be reduced. This can be represented by an IF–THEN rule

IF (the frictional power loss > the design criterion)
THEN (decrease the friction coefficient).

In addition, in the intelligent system for tribological design of engine, the rules can be used to represent domain experts' design strategies. For example

> **IF** (the frictional power loss > the design criterion)
> **THEN** (decrease the number of piston rings, or decrease width of piston ring.)

If there are enough IF–THEN rules to describe each inference step of the design procedure, then the design decision will be a direct and automatic inference process. If all the design parameters and variables referred to during the design process can be described as symbol models, and all the design strategies can be described step-by-step using the rules from experience and knowledge, then the design problem is assumed to be a "Normal Design Problem" in our nomenclature.

Besides rules, framework and object-oriented models can also be used in engine tribological design. Taken together, they make up an explicit knowledge representative form, which corresponds with a serial inference process, and which reflects the most important and popular thought modes of engine designers. A symbol model-based intelligent system for engine tribological design has already been implemented, and this will be discussed in detail in Chapter 3.

Artificial neural network model

Engine tribological design is a complicated design procedure that needs various types of information or knowledge. Some of them can be represented as explicit symbol models. However, other are implied in the abundant experimental data or numerical calculation data, and cannot be directly expressed even by domain experts. In this case, an artificial neural network (ANN) model is available to simulate the implicit complicated functional relationships included in the abundant engineering data, and then to extract valuable knowledge from them.

Modern biology has demonstrated that there are about 10^{11} brain cells in the human brain. Each brain cell is an active information-processing unit, called a neural unit. Each neural unit is connected to another in about 10^5 arcs. All the neural units and their connecting arcs make up a complicated net, called a human brain neural net. Because of this net, human beings can perform complex thought processes, and make judgments immediately when faced with complicated situations.

Artificial neural networks are massive parallel networks of simple processing elements designed to emulate the function and structure of the human brain. According to the structure of the human brain's neural net, a simple but typical artificial neural network can be built consisting of an input layer and an output layer, such as the net illustrated in Fig.1-10.

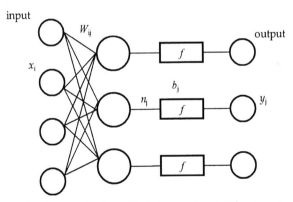

Fig.1-10. A simple, but typical, artificial neural net. The input and output layers are composed of several neural units. Each layer owns its trigger function, f, which represents the functions of the neural unit.

In the ANNs, each layer is composed of several neural units, and each neural unit has an internal trigger function, f, and an offset, b. The trigger function is the core of a neural unit, which is activated only when the input value is "TRUE". Usually, the trigger functions in any one layer are the same, the trigger function in an input layer is a linear function, and the trigger function in an output layer is a Sigmoidal function as follows[16-17]

$$f(n+b) = \frac{1}{1+\exp[-(n+b)]} \qquad (1-2)$$

where n is the input sum of the neural unit. According to the ANN model in Fig.1-10, an input neural unit, x_i, is connected to an output neural unit, y_i, by its connecting arc with a weight, W_{ij}. Weights reflect the connection strength between the units. If a connecting arc does not exist, then its weight is set to zero. Then, the input sum of y_j neural units is

$$n_j = \sum_{i=1}^{m} W_{ij} x_i \qquad (1-3)$$

where m is the number of neural units in the input layer. Then, the output of the y_j neural units is

$$y_j = f(n_j + b_j) \qquad (1-4)$$

The output of the entire network is given by

$$A = \{y_1, y_2, y_3 \ldots \ldots y_n\} \qquad (1-5)$$

From its structure (the units and arcs) and trigger functions, an ANN can represent complicated non-linear mapping relationships between its input layer parameters and its output layer parameters. Different functional relationships may correspond with different structures and weights of the neural net.

ANNs can learn from experience and by previous examples. They modify their structure and weights in response to the environment. The behaviour of the trained ANN depends on its structure and weights. To decide the structure and weights of an ANN for a practical problem, a group of training samples needs to be built, each should contain input data and the corresponding output data. After the ANN has been trained, the complicated functional relationships implied in the training samples are transferred into a special structure, and corresponding the arc weights of the ANN can be further extracted and transformed into explicit knowledge forms if needed[11].

One of the distinct features of the ANN is its ability to learn from experience and examples and then to adapt with changing situation. This learning process occurs even when the input data contains errors or is incomplete or fuzzy, which is often typical of a design process. This characteristic of ANN makes it a promising candidate for modelling some of the engineering problems.

An interesting engineering definition of ANN is, "A computational mechanism able to acquire, represent and compute mapping from one multivariate space of information to another, given a set of data representing that mapping."

In engineering practice, abundant calculation and experimental data exist, which are extremely valuable for design decision-making. As they can be formed into training samples easily, ANN can be adopted as a knowledge model to represent the implied complicated functional relationships for extracting valuable implicit knowledge. After the ANN has been trained, it is able to generate rules and will be able to respond within the domain covered by the training samples.

For an engine triblogical design, ANN are extremely useful, especially in the situation for which rules are either not know or very difficult to discover. Some of the major attributes of ANN used in engine tribological design can be listed as:

(1) Data presented to training ANN can be theoretical data, experimental data, and empirical data based on good and reliable past experience or a combination of these.

(2) ANN can learn and generalize from data or examples to produce meaningful solutions to problems.
(3) ANN can perfectly cope with situation where the input data to the network is fuzzy, discontinuous or is incomplete.
(4) ANN is able to adapt solutions over time and to compensate for changing circumstances.

These attributes make ANN a promising tool for modelling knowledge related to tribological design in engines. For example, in the tribological design of an piston assembly, ANN can be used to simulate the complicated functional relationships between the frictional power loss and the piston design parameters, which are implied in the abundant calculation data. Furthermore, rules can be extracted from the trained ANN model, which can then be used to instruct the design procedure. The ANN model, and its newly extracted rules, can make up the shortage of symbol models in dealing with rule conflict problems. The details of an ANN model and its application are described in Chapter 4 of this book.

Gene model

Symbol models can represent the domain experts' explicit knowledge and transform design decisions into a serial inference procedure. ANN can simulate complicated functional relationships or the implicit knowledge implied in experimental or calculation data, and can transform these into rules to expand the coverage of symbol models. The previously defined concept, normal design problem, can be dealt with directly and efficiently by an intelligent system based on the symbol and the ANN models.

However, not all of the design problems encountered in tribological design of engines are normal design problems, because they are essentially multi-purpose, multi-constraint, and multi-parameter problems. During the design process, a serial inference mechanism is insufficient to deal with the situation where several design purposes or constraints run in parallel or in conflict. This is the situation a comprehensive parallel decision support mechanism is required. Thus, a new parallel knowledge model needs to be adopted, called the gene model.

The gene model derives from the concept involved in genetic algorithms (GAs). GAs are robust search mechanisms based on the principle of population genetics, natural selection, and evolution[18-20].

It is well known that, the basic structure and features of living organisms are determined by their gene arrangements (*i.e.*, chromosomes), which can be reproduced, crossed, or mutated to enable the creature population to propagate, evolve, or degenerate. This process obeys to the principles of natural selection and the "survival of the fittest", *i.e.*, the fittest creatures in a population have more opportunities to survive and reproduce new offspring than others. All the new offspring make up a new generation, which will evolve continually under the above principles.

The GAs was first developed as an optimization algorithm by John Holland and his colleagues and students at Michigan University[18]. Genetic algorithms simulate the evolutionary mechanism of living organisms. According to their mechanism, a potential solution to an optimization problem can be encoded into a binary string (*i.e.*, a gene string), in which each individual bit is called a gene. Therefore, a group of gene strings make up a potential solution set, namely solution population. GAs search the space composed of the possible solutions for optimum individuals. Each potential solution is assigned to a set of fitness function value by the "environment" to determine how well it is in terms of solving the problem. The fitness function value determines its probability of surviving to the next generation. Various GA gene operators function on the solution population to identify better solutions, to eliminate undesirable solutions, and to determine whose offspring will survive to the next generation. The evolutionary process continues until optimum solutions are achieved. Obviously, GAs are population-based iterative optimization searching procedure, maintaining a population of candidate gene samples over many generations. The entire search process is guided by the values of the fitness function, which mimic the living selection standards of nature. During the process, every gene string in the population is functioned by the various genetic operators, which mimic the evolutionary mechanism exists in nature.

Generally, GAs have four components: encoding mechanism, fitness function, gene operators, and controlling parameters[19,20]. In the ensuing discussion, these will be illustrated in terms of engineering design problems, in which the potential design solutions are represented as gene strings.

Encoding mechanism. In nature, the inherited characteristics of each organism are contained in its chromosomes (gene strings). Different gene strings correspond to different individuals, and an individual can be identified from its chromosomes definitely. Accordingly, in the design field, a scheme made up of several design parameters and their values can be encoded into a binary gene string, called a gene sample. In the gene sample, a binary gene segment corresponds to a design parameter, its value corresponds to the value of the design parameter. Then, a group of gene samples forms a design solution population, in which each gene sample corresponds to a unique design scheme.

Fitness function. This reflects the functional relationship between a design scheme (*i.e.*, a gene sample) and its corresponding design objectives and constraints. Generally, the function value, f, denoting the fitness value, is scaled between 0 and 1 to evaluate different design samples. To calculate f, the GAs must first decode the binary representation of the design scheme maintained in a gene sample, and then analyse its performances and calculate its corresponding fitness value. Based on their various *f* values, the design schemes in a population can be classified into fitter samples, bad samples and so on.

Gene operators. After the gene samples in the current generation have been evaluated according to their fitness values, f_i, the gene operators are then applied to propagate the fitter samples and eliminate bad samples. The most important and essential gene operators in a GA are the reproduction operator, mutation operator, and crossover operator.

The reproduction operator copies gene samples to the next generation according a roulette wheel mechanism, in which the probability of a sample being selected, the reproduction rate, P_r, is proportional to its fitness value

$$P_r = f_i \Big/ \sum f_i \qquad (1\text{-}6)$$

where f_i is the fitness value of the operated gene sample, and $\sum f_i$ is the sum of the fitness values over the whole population. From Equation (1-6), it can be seen that samples with higher fitness value will have more opportunities to be selected and copied to the next generation. Under the application of the reproduction operator, the population can evolve, while improving the average quality of the population.

The crossover operator produces new offspring by swapping gene segments of two parent samples. It selects a certain proportion of individuals within a population according to the crossover rate, P_c, and swaps them with fitter samples at a crossover site to generate new offspring, as shown in Fig. 1-11.

The mutation operator randomly selects a gene sample according to the mutation rate, P_m, and randomly alternates its individual gene bit from 1 to 0, or vice versa, as shown in Fig. 1-12. Under the function of the two operators, new gene samples are generated and kept in the next generation, thus the diversity of the new population is maintained.

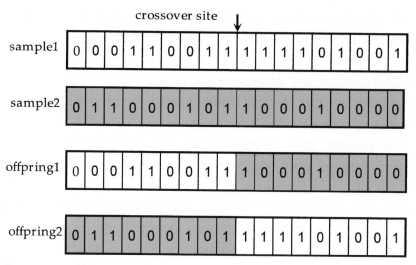

Fig.1-11 An illustration of the crossover operator. Two samples are swapped at a crossover site to generate new offspring for the next generation.

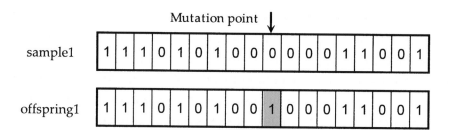

Fig.1-12 An illustration of the mutation operator. A gene in a sample is mutated from 0 to 1. Then the new offspring is kept in the next generation. The mutation operator can maintain the diversity of a population to avoid local convergence.

Control parameters. The control parameters in GAs include population size, crossover rate, mutation rate, and termination criterion. The population size, n, is the number of gene samples in the population of the current generation. The crossover rate, P_c, and the mutation rate, P_m, indicate the ratios of the number of operated samples to the population size, n. The termination criterion is defined as a given tolerance or as a predefined maximum number of completed generations that determine the convergence of the GAs. These parameters control the iterative evolutionary process of GAs until the optimum design solution has been reached. All of them have great influence on the results of optimisation, but what values should be selected is still a matter of trial and error. Generally, the values chosen depend on the type of problem and the kind of GAs used.

Population size is one of the most important parameters. If the population s1ze is too small, the GAs may converge to a local minimum. If the population size is too large, the GAs may waste computational resources, which means that the time for an improvement is very long. For the GAs with a constant population size, the most common sizes of population vary from 50 individuals to 500. For a large population (*e.g.*, n is lager than 100), P_c is set to around 0.06 and P_m set to around 0.01. For a small population size (*e.g.*, n is around 30), P_c is set to around 0.08, and P_m set to around 0.04.

When used in engine tribological design problems, GAs encode the potential design solutions into binary gene strings and form a gene population. The GAs then integrate the design problem's multiple purposes, multiple constraints, and multiple parameters into a fitness function. At each evolutionary step (*i.e.*, generation), each individual of the current population is decoded and evaluated to obtain its fitness function value. Then, various gene operators are triggered to evolve the population. The procedure is repeated to search out optimum and feasible design solutions automatically and concurrently. Therefore, although it is time-consuming, gene string is a powerful knowledge model for dealing with comprehensive parallel decision problems, especially when there is a shortage of domain experiential knowledge. Their potential applications in tribological design of engines will be investigated in Chapters 5 and 6 of this book.

1.5 Characteristics of the three types of knowledge models

As mentioned above, the knowledge models proposed in this book reflect the thought modes of human beings, or the evolutionary mechanisms found in nature. When used in engineering design, each of them has its advantages and disadvantages, and each is suitable for dealing with different design problems.

Meanwhile, the tribological design of engines is a very complicated design problem that involves numerical analysis, rule-based serial inference, rule conflict solving, parallel inference, and comprehensive decision-making. Therefore, to built an intelligent system for tribological design of engines, it is necessary to integrate all the three types of knowledge models to take good advantage of them. The characteristics of the three types of knowledge models are listed in Table 1-2.

1.6 Summary

The tribological design methodology for an internal combustion engine is suggested to built using an intelligent computer aided design system, which aims at combining domain experiential knowledge with tribological theoretical analysis programs to directly and effectively deal with design problems. The core function of such an intelligent system is its decision-making mechanism and intelligent ability.

Three types of knowledge models are proposed to construct the intelligent system: the symbol model, the artificial neutral network model, and the gene model. Different knowledge models are suitable for dealing with different design problems. The ICAD system for engine tribological design should integrate all the three knowledge models and so take good advantage of each.

Table1-2. *The characteristics of knowledge models*

characteristics / model	Origin	Representation mechanism	Application mechanism	Application process	Speed
Symbol model	Human thought mode	Rule, framework, object-oriented models	Serial inference mode	Transparent	Quick, Direct.
Artificial neutral network model	Information processing mechanism in human brain neutral networks	Structure and connecting weights of ANN	Extract valuable rules for design	Non-transparent	Medium
Gene model	Simulating genetic and evolutionary mechanism found in nature	Chromosome (gene sample), fitness function, and operators	Comprehensive, parallel decision-making	Semi-transparent	Slow, time-consuming

References

1. J. Halling. Principles of tribology. Macmillan Press, 1975
2. T. E. Tallian. Tribological design decisions using computerized databases. *Transactions of ASME: Journal of Tribology,* 1987,109(3), pp381-387
3. C. M. Taylor. Automobile engine Tribology - design considerations for efficiency and durability. Wear, 1998, 221, pp 1–8
4. C. L. Gui. A study on tribological design methods for machine parts. *Chinese Journal of Mechanical Engineering.* 1992, vol. 5(4), pp 268–274
5. C. M. Taylor. Engine Tribology. Tribology Series: 26, Elsevier Science Publishers B.V. 1993
6. D. Dowson, C. M. Taylor, M. Godet, and D. Berthe. Tribology of reciprocating engines. *Proceedings of the Ninth Leeds-Lyon Symposium on Tribology.* Butterworth and Co, 1983
7. D. Dowson, C. M. Taylor, and M. Godet (Eds). Vehicle tribology. Tribology series, 18, Elsevier Science Publishers, 1991
8. D. R. Adams. Design and analysis: A perspective for the future. Vehicle Tribology, D. Dowson, C. M. Taylor and M. Godet (Eds.), Tribology series, 18, Elsevier Science Publishers, 1991, pp 7–15
9. V. Akman, P. J. W. Ten Hagen, and T. Tomiyama. A fundamental and theoretical framework for an intelligent CAD system. *Computer-Aided Design.* 1990, vol. 22, No.6, pp 352-67
10. M. Burry, J. Couslon, J. Preston, et al. Computer-aided design decision support: Interfacing knowledge and information. *Automation in Construction.* 2000,10(2), pp 203-215
11. S. Tzafestas (Ed.). Expert systems in engineering applications. Berlin: Springer-Verlag, 1993
12. C. Shakeri, I. Deif, P. Katragadda, et al.. Intelligent design system for design automation. *Intelligent systems in Design and manufacturing III. B.* Gopalakrishnan and A. Gunasekaran (Eds.), Boston: SPLE, 2000, pp 27–35

13. F. Kimura. Architecture and implementation subgroup—summary of the workshop discussion. *Intelligent CAD, III.*, H. Yoshikawa, F. Arbab, and T. Tomoyama (Eds.) North-Holland: Elsevier Science Publishers B.V., 1991, pp 43–49

14. M. Nagao. Possible contributions of information processing technology to CAD. *Intelligent CAD, III.*, H. Yoshikawa, F. Arbab, and T. Tomoyama (Eds.) Elsevier Science Publishers B.V. (North-Holland), 1991, pp 3–10

15. X. J. Zhang. Study on the design methods of tribological design for internal combustion engine. PhD. dissertation, Hefei University of Technology,PRC, Dec. 2000

16. A. Hojjat. Knowledge engineering: fundamentals. Springer Verlag,1990

17. A. S. Garcez, K. Broda, and D. M. Gabbay. Symbolic knowledge extraction from trained neural networks: A sound approach. *Artificial Intelligence.* Vol.125, 2001:155–207

18. D. A. Coley. An introduction to genetic algorithms for scientists and engineers. World scientific Publishing, Singapore, 1999

19. D. E. Goldberg. Genetic Algorithms in Search, Optimization, and Machine Learning, AWPC,1989

20. M. A. Rosenman. An exploration into evolutionary models for non-routine design. *Artificial Intelligence in Engineering*, 1997, vol. 11, pp 287–293

Chapter 2: General Consideration of Tribological Design of an Engine

As mentioned above, from the viewpoint of tribology, an internal combustion engine is composed of several subsystems including the bearing system, piston system, and the valve train system. To predict and design the tribological performances of an engine, it is essential to establish an analytical or experimental relationship between the design aims and design elements. Therefore, tribological design principles for an engine are will be presented in this chapter.

2.1 Tribological design for the bearing system of an engine

2.1.1 General consideration of design principles

The bearing friction pairs of an engine include main bearings of the crankshaft, big-end bearings and small-end bearings of the connecting rod. The lubrication conditions and analysis methods of the three kinds of bearings are quite different due to their different loading features and movement characteristics. Lubrication conditions of the main bearings also vary from each other greatly as a result of the different operating sequences of engine cylinders. Meanwhile, since the bearing system carries a transient and dynamic load varying greatly, its physical dimension and clearance need to be highly restricted. The strictly restricted dimension and clearance make the minimum film thickness of a bearing at the same level as the surface roughness, which is apt to give rise to the scratching of a bearing. If a bearing scratch occurs, the result will be terrible in engineering practice.

In order to prevent the engine bearing system from scratching and meanwhile to decrease its frictional power loss, researchers on engine tribology have carried out extensive and profound investigation into the friction pairs of a bearing system. There are more and more actual and effective influencing factors being gradually considered into the lubrication analysis and calculation of an engine bearing system. The relevant research work may be categorized into several aspects, which are introduced as follows accompanied with the introduction to the basic tribological design principles adopted in this book.

(1) Calculation of the journal centre orbits

Methods employed to calculate the centre orbits of bearing include Hahn's approach proposed by H.W. Hahn[1], Holland's approach by J. Holland[2], and Mobility approach by J.F. Booker[3]. The principles of Hahn's approach are described as follows. Firstly, the dynamically loaded Reynolds equation was divided into two parts, i.e., rotational motion and squeeze motion. Based on unified boundary conditions, the pressure distribution of oil film was resolved using linear additive principles. Then, the bearing load force was obtained by vector composition, and finally the journal centre orbits were calculated by solving the static equilibration equation between the whole cycle of bearing load and the external applied load. This approach was quite strict in mathematics, but its calculation amount was very huge. In order to overcome the difficulties in solving the Reynolds equation, Holland calculated the rotational motion and squeeze motion according to their independent boundary conditions respectively. Then the rotational oil film force and the squeezing oil film force were equilibrated with the external applied load. Finally, motion equations were derived and then solved to obtain the centre orbits of bearing. This method neglected the interrelationship between the two motions (rotation and squeeze motion) and their effects on the whole load capacity, and the relevant calculation equations were still significantly complex. However, the calculation process was simple, and the calculation amount was reduced. Therefore, this method was practical in the design of an engine bearing system under normal demands. The Mobility method was based on the infinite narrow bearing theory, so the analytic solution of oil film pressure could be obtained. Both the solving speed and the accuracy were quite satisfying, therefore, it found a widespread application in the bearing design. However, since the infinite narrow assumption was adopted and the oil feed features were neglected, the Mobility method was not available in precise bearing analysis. Based on the calculation ability of the intelligent system, Hahn's method was adopted in this book.

(2) Lubrication analysis considering surface roughness

Since the surface roughness of an engine bearing is at the same level as the minimum film thickness, the effect of surface roughness on the tribological properties of a bearing system is non-negligible. Based on Christensen random model, Prof. Qiu Zugan[4] analysed the 1-dimensional roughness lubrication of a finite length bearing with dynamic load. Combining the average flow model and Hahn's method, Dr. Wang Xiaoli[5] established a numerical analysis approach for 2-dimensional surface roughness lubrication. According to the expert knowledge sorted out in this book, Dr. Wang's approach is suitable to the bearing system of an engine with power output per litre less than 11KW/L, strengthening coefficient less than 6MPa.m/s and average cylinder effective pressure less than 0.7MPa.

(3) Lubrication analysis considering thermal effects

As mentioned above, working conditions of an engine bearing system are severely tough, which easily give rise to terrible scratching. Meanwhile, scratching is always associated with heat. Therefore, for the internal combustion engine of a vehicle like an automobile, lubrication analysis of a bearing system considering thermal effects, i.e. Thermohydrodynamic (THD) Lubracation analysis, is indispensable. However, the unsteady periodic external load acting on the bearing system makes it very difficult to proceed THD analysis or calculation. When the thermal effects are also considered, it is needed to solve an energy equation simultaneously, which will result in an extremely large calculation. In 1995, R.S. Paranjpe[6] obtained a complete numerical solution for dynamically loaded bearing for the first time. In his analysis, journal was assumed as an isothermal body, and the thermal effects of bush, journal and lubricant film were considered, which were different and varied with time. In 1999, Dr. Wang Xiaoli[7] proposed a THD analysis model for main bearing system considering the surface roughness effect, and obtained a complete numerical solution. The THD analysis model aimed to determine the journal centre orbits during a loading cycle. In order to determine a transient location of the journal centre, a group of simultaneous equations needed to be solved, which included the average Reynolds equation considering 2-dimensional surface roughness, energy equation, heat conduction equation and load equilibrium equation. It has been demonstrated that, this analysis model is suitable for the internal combustion engine with high strengthening coefficient.

(4) Lubrication analysis considering the oil feed features

Eliminating the infinite narrow assumption in Mobolity method, J.G. Jones[8] considered the effect of oil feed grooves on the loading capacity of oil film, and adopted an oil film history model. The calculation results were in good agreement with the actual measurement results. However, the calculation process was heavily time-consuming, nearly 10 thousand times slower than that of Mobility methods for short bearings. Therefore, it was difficult to be a rapid design method for a bearing system. Basing on finite element method, K.P. Goenka[9] obtained the finite length bearing solution. He especially investigated the influence on the minimum film thickness of the circumferential or longitudinal irregular geometry including partial oil groove, complete oil groove, conical or axial asymmetrical groove. His research indicated that, even a slightly irregular geometry would have a significant effect on the minimum film thickness. Goenka's analysis method was more accurate and faster than Jones's method, however, its calculation time was still un-acceptable for engineering design. In this book, two analysis approaches by Dr. Wang

Xiaoli are adopted, i.e., the isothermal lubrication analysis of engine bearing considering the surface roughness effect, the thermohydrodynamic lubrication analysis of engine bearing considering the surface roughness effect. Both of them are used to investigate the effect of oil feed features on the loading capacity. The isothermal lubrication analysis indicated that the oil film pressure hold a minimum value for complete oil groove, a maximum value for no existing of oil groove, and an intermediate value for a partial oil groove. The thermohydrodynamic lubrication analysis indicated that oil groove resulted in an obvious decrease in film pressure and film temperature, therefore, would reduce the loading capacity of oil film, but would benefit bearing cooling.

In addition to the four aspects mentioned above, research on lubrication theory and analysis of engine bearing includes lubrication analysis considering the non-Newtonian properties of lubricant and the cavity boundary conditions, as well as the lubrication analysis of connecting rod bearing considering the elastic deformation. However , these theories and analysis are not adopted in this book, because the current calculation ability can not deal with so many non-linear equations simultaneously. As we proceed with our future research, we expect to simultaneously bring more and more actual influencing factors into the engine bearing lubrication analysis

2.1.2 Mathematic equations for design and calculation

2.1.2.1 Isothermal lubrication analysis considering surface roughness effect

The loading capacity of oil film, lubricant flow, frictional power loss and bearing temperature may be obtained by considering a simultaneous solution of a 2-dimensional average Reynolds equation with the viscosity-temperature and density-temperature properties of lubricant.

During the calculation, the pressure flow factors and shear flow factors of the average Reynolds equation are required to be determined in advance.

Dimensionless average Reynolds equation and its boundary conditions

For a journal bearing with rough surface as shown in Fig.2-1, an average Reynolds equation proposed by Patir and Cheng[10] is,

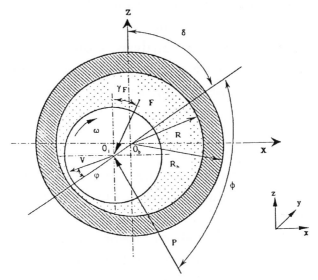

Fig. 2-1   the coordinate system of a bearing system

$$\frac{\partial}{\partial x}\left(\phi_x\frac{h^3}{12\eta}\frac{\partial p_T}{\partial x}\right)+\frac{\partial}{\partial y}\left(\phi_y\frac{h^3}{12\eta}\frac{\partial p_T}{\partial y}\right)=\frac{U_1+U_2}{2}\frac{\partial h_T}{\partial x}+\frac{U_1-U_2}{2}\sigma\frac{\partial\phi_s}{\partial x}+\frac{\partial h_T}{\partial t}$$

$$(2-1)$$

where, ϕ_x,ϕ_y — pressure flow factors in the x, y direction respectively

ϕ_s — shear flow factor

h — nominal film thickness

h_T — average film thickness or average gap

p_T — mean hydrodynamic pressure

η — absolute kinematic viscosity

t — time

U_1, U_2 —linear speed of journal and bush respectively

σ — standard deviation of composite surface roughness

The average gap between the two rough surfaces of bearing is,

$$h_T = \frac{h}{2}\left[1 + erf\left(\frac{h}{\sqrt{2}\sigma}\right)\right] + \frac{\sigma}{\sqrt{2\pi}}e^{-h^2/2\sigma^2} \tag{2-2}$$

where $erf()$ is an error function.

For a crankshaft main bearing, the bush is stationary and the journal is rotational, therefore $U_2 = 0, U_1 = U$. Combining equations (2-1) and (2-2), and leading in the following dimensionless parameters,

$$\theta = x/R, \bar{y} = y/(B/2), \bar{h} = 1 + \varepsilon\cos\theta, \bar{h} = h/C,$$

$$\bar{p} = pc^2/\eta\omega R^2, \omega = U/R, \bar{t} = \omega t, \Lambda = c/\sigma$$

where, R —bearing radius

$\quad B$ — bearing width

$\quad \varepsilon$ — bearing eccentricity ratio

$\quad c$ — radial clearance of bear

$\quad \omega$ — rotational velocity of journal

$\quad \Lambda$ — Lambda ratio, i.e., the ratio of film thickness to composite surface roughness

Substituted by these parameters, a dimensionless average Reynolds equation for the main bearing can be obtained,

$$\frac{\partial}{\partial\theta}(\phi_x\bar{h}^3\frac{\partial\bar{p}}{\partial\theta}) + \left(\frac{1}{\lambda}\right)^2\frac{\partial}{\partial\bar{y}}(\phi_y\bar{h}^3\frac{\partial\bar{p}}{\partial\bar{y}})$$

$$= -3\varepsilon\sin\theta(1 - \frac{2}{\omega}\frac{d\delta}{dt})\left[1 + erf(\frac{\Lambda\bar{h}}{\sqrt{2}})\right] + \frac{1}{\Lambda}\frac{\partial\phi_s}{\partial\theta}$$

$$+6\cos\theta\frac{1}{\omega}\frac{d\varepsilon}{dt}\left[1 + erf(\frac{\Lambda\bar{h}}{\sqrt{2}})\right] \tag{2-3}$$

where, λ — width-diameter ratio of bearing

Now based on the principles of Hahn's method and the linear additive principle, equation (2-3) can be divided into the following three equations,

$$\frac{\partial}{\partial\theta}(\phi_x\bar{h}^3\frac{\partial\bar{p}_1}{\partial\theta})+\left(\frac{1}{\lambda}\right)^2\frac{\partial}{\partial\bar{y}}(\phi_y\bar{h}^3\frac{\partial\bar{p}}{\partial\bar{y}})=-3\varepsilon\sin\theta\left[1+\mathrm{erf}(\frac{\Lambda\bar{h}}{\sqrt{2}})\right] \qquad (2\text{-}4)$$

$$\frac{\partial}{\partial\theta}(\phi_x\bar{h}^3\frac{\partial\bar{p}_2}{\partial\theta})+\left(\frac{1}{\lambda}\right)^2\frac{\partial}{\partial\bar{y}}(\phi_y\bar{h}^3\frac{\partial\bar{p}_2}{\partial\bar{y}})=6\cos\theta\left[1+\mathrm{erf}(\frac{\Lambda\bar{h}}{\sqrt{2}})\right] \qquad (2\text{-}5)$$

$$\frac{\partial}{\partial\theta}(\phi_x\bar{h}^3\frac{\partial\bar{p}_3}{\partial\theta})+\left(\frac{1}{\lambda}\right)^2\frac{\partial}{\partial\bar{y}}(\phi_y\bar{h}^3\frac{\partial\bar{p}_3}{\partial\bar{y}})=\frac{1}{\Lambda}\frac{\partial\phi_s}{\partial\theta} \qquad (2\text{-}6)$$

The boundary conditions are,

(1) periodic boundary condition: $\bar{p}_{1,2,3}\big|_{\theta=0}=\bar{p}_{1,2,3}\big|_{\theta=2\pi}$

(2) oil groove boundary condition: $\bar{p}\big|_{\Gamma}=\bar{p}_s$

(3) end face boundary condition: $\bar{p}_{1,2,3}\big|_{\bar{y}=\pm1}=0$

(4) symmetric boundary condition: $\dfrac{\partial\bar{p}_{1,2,3}}{\partial\bar{y}}\bigg|_{\bar{y}=0}=0$

where Γ — oil groove boundary, $\bar{p}_s$ — pressure of the feeding oil

Let $q_1=1-\dfrac{2}{\omega}\dfrac{d\delta}{dt}, q_2=\dfrac{1}{\omega}\dfrac{d\varepsilon}{dt}$, then the dimensionless oil film pressure $\bar{p}$ can be represented as,

$$\bar{p}=q_1\bar{p}_1+q_2\bar{p}_2+\bar{p}_3 \qquad (2\text{-}7)$$

At the boundary of oil film breakage where $\theta=\theta_s$, the Reynolds boundary condition is satisfied, i.e., $\dfrac{\partial\bar{p}}{\partial\theta}\bigg|_{\theta=\theta_s}=0, \bar{p}\big|_{\theta=\theta_s}=0.$

Viscosity-temperature relationship

The Vogel model was adopted to express the viscosity-temperature relationship of lubricant,

$$\eta = ae^{b/(T+c)} \tag{2-8}$$

where, a, b, c —lubricant coefficients determined by experiments[11].

Density-temperature relationship

In our research, the density-temperature relationship may be determined from *the conversion table of density experiment and measurement* of China[12]. For example, based on the lubricant density ρ_{20} measured under 20℃, coefficient β can be found from the conversion table, then density ρ_T under different temperature T can be calculated out using equation,

$$\rho_T = \rho_{20} - \beta(T - 20) \tag{2-9}$$

Loading capacity of oil film, lubricant flow, frictional power loss and bearing temperature

(1) Loading capacity of oil film

The dimensionless loading force of oil film $\overline{p}$ has a parallel component $\overline{p}_p$, and a vertical component $\overline{p}_v$, as follows,

$$\overline{p}_p = -\int_A \overline{p} \cos\theta dA, \qquad \overline{p}_v = -\int \overline{p} \sin\theta dA \tag{2-10}$$

Therefore, dimensionless loading force $\overline{p}$ and its acting angle φ are,

$$\overline{p} = \sqrt{\overline{p}_p^2 + \overline{p}_v^2}, \qquad \varphi = \arcsin\left(\frac{\overline{p}_v}{\overline{p}}\right) \tag{2-11}$$

(2) Lubricant flow

The transient lubricant flow Q_α is

$$Q_\alpha = -\frac{R^3\psi}{6\lambda}\omega\int_{\theta_1}^{\theta_2}\phi_y(1+\varepsilon\cos\theta)^3\left(\frac{\partial p}{\partial y}\right)_{\overline{y}=1}d\theta \tag{2-12}$$

where, ψ — relative clearance, $\psi = c/R$;

λ — width-diameter ratio of the bearing

θ_1, θ_2 — starting and terminal angles of oil film

For a 4-stroke diesel engine, average lubricant flow in a loading cycle is,

$$Q_m = \frac{1}{2\pi} \int_0^{2\pi} Q_\alpha d\alpha \qquad (2\text{-}13)$$

(3) Frictional power loss

The transient frictional power loss is[13],

$$N_{f\alpha} = \frac{BR^2 \eta \omega^2}{\varphi} \left[\int_{\theta_1}^{\theta_2} \frac{\phi_f - \phi_{fs}}{1 + \varepsilon \cos \theta} d\theta + \right.$$

$$\left. \frac{\varepsilon}{2} \int_0^1 \int_{\theta_1}^{\theta_2} \phi_{fp} \, \bar{p} \sin \theta d\theta d\bar{y} + (1 + \varepsilon \cos \theta_2) \int_{\theta_1}^{\theta_2} \frac{\phi_f - \phi_{fs}}{(1 + \varepsilon \cos \theta)^2} d\theta \right] \qquad (2\text{-}14)$$

where $\phi_f, \phi_{fs}, \phi_{fp}$ — shear stress factors, which will be determined in

the following section of this chapter.

For a 4-stroke diesel engine, average frictional power loss in a loading

circle (2π) is,

$$N_m = \frac{1}{2\pi} \int_0^{2\pi} N_{f\alpha} d\alpha \qquad (2\text{-}15)$$

(4) Average temperature of bearing

It is assumed that, during the flow process of lubricant, the heat resulting from the viscous dissipation is completely absorbed by the lubricant, and then the oil flows out from the two ends of bearing[14]. Therefore, there exists a thermal equilibrium equation,

$$N_{f\alpha} = c_f \rho Q_\alpha (T_\alpha - T_0)$$

So the transient temperature is

$$T_\alpha = \frac{N_{f\alpha}}{C_f \rho Q_\alpha} + T_0 \qquad (2\text{-}16)$$

where c_f — specific heat of lubricant under constant pressure

ρ — density of lubricant

T_0 — temperature of the feeding oil

T_α — transient temperature of the outflow oil at the two ends of bearing

For a 4-stroke diesel engine, the average temperature in a loading circle (2π) is,

$$T_m = \frac{1}{2\pi} \int_0^{2\pi} T_\alpha d\alpha \qquad (2\text{-}17)$$

Approximate temperature of the bearing is,

$$T = \frac{1}{2}(T_0 + T_m) \qquad (2\text{-}18)$$

Determining of the flow factors and shear force factors

(1) Pressure flow factors are determined by follows[10]:

$$\phi_x = 1 - c_1 e^{-g\Lambda \bar{h}} \quad \text{for } (\gamma \le 1)$$

$$\phi_x = 1 + c_1(\Lambda \bar{h})^{-g} \quad \text{for } (\gamma > 1)$$

$$\phi_y(\Lambda \bar{h}, \gamma) = \phi_x(\Lambda \bar{h}, 1/\gamma) \qquad (2\text{-}19)$$

where, γ - surface direction parameter. $\gamma < 1$ means a cross directional texture surface, $\gamma = 1$ means an isotropic texture surface, and $\gamma > 1$ means a longitudinal directional texture surface. c_1, g are constants referred to Table2-1.

Table 2-1. the values of c_1 and g

γ	c_1	g	Range
1/9	1.48	0.42	$\Lambda \bar{h} > 1$
1/6	1.38	0.42	$\Lambda \bar{h} > 1$
1/3	1.18	0.42	$\Lambda \bar{h} > 0.75$
1	0.9	0.56	$\Lambda \bar{h} > 0.5$
3	0.225	1.5	$\Lambda \bar{h} > 0.5$
6	0.520	1.5	$\Lambda \bar{h} > 0.5$
9	0.870	1.5	$\Lambda \bar{h} > 0.5$

(2) Shear flow factors are determined by follows[10],

$$\phi_s = \left(V_{rj} - V_{rb}\right)\Phi_s = \left(2V_{rj} - 1\right)\Phi_s \tag{2-20}$$

where, $\Phi_s = A_1\left(\Lambda\bar{h}\right)^{\alpha_1} e^{-\alpha_2\Lambda\bar{h}+\alpha_3\left(\Lambda\bar{h}\right)^2}$ for $\Lambda\bar{h} \le 5$;

$\Phi_s = A_2 e^{-0.25\Lambda\bar{h}}$ for $\Lambda\bar{h} > 5$.

V_{rj}, V_{rb} — surface roughness variance ratios of journal and bush respectively, $V_{rj} = \left(\sigma_j/\sigma\right)^2, V_{rb} = \left(\sigma_b/\sigma\right)^2$. $V_{rj} = 0$ denotes a completely smooth journal and a rough bush; $V_{rj} = 1$ denotes a completely smooth bush and a rough journal; $V_{rj} = 0.5$ denotes that journal and bush are of the same roughness.

σ_j, σ_b — root-mean-square values of journal and bush roughness. Values of $A_1, A_2, \alpha_1, \alpha_2, \alpha_3$ are listed in Table2-2.

Table 2-2. the values of $A_1, A_2, \alpha_1, \alpha_2, \alpha_3$

γ	A_1	A_2	α_1	α_2	α_3
1/9	2.046	1.856	1.12	0.78	0.3
1/6	1.962	1.754	1.08	0.77	0.3
1/3	1.858	1.561	1.01	0.76	0.3
1	1.899	1.126	0.98	0.92	0.5
3	1.560	0.556	0.85	1.13	0.8
6	1.290	0.388	0.62	1.09	0.8
9	1.011	0.295	0.54	1.07	0.8

(3) Shear stress factors are determined by the following equations[10],

$$\phi_f = \frac{35}{32}z\left\{\left(1-z^2\right)^3 \ln\frac{z+1}{\xi^*} + \frac{1}{60}\left[-55 + z\left(132 + z\left(345 + z\left(-160 + z\left(-405 + z\left(60 + 147z\right)\right)\right)\right)\right)\right]\right\}$$

for $\Lambda\bar{h} \le 3$

$$\phi_f = \frac{35}{32}z\left\{\left(1-z^2\right)^3 \ln\frac{z+1}{z-1} + \frac{z}{15}\left[66 + z^2\left(30z^2 - 80\right)\right]\right\}$$

for $\Lambda\bar{h} > 3$

$$\tag{2-21}$$

where $z = \Lambda\bar{h}, \xi^* = 0.00333$

$$\phi_{fp} = 1 - De^{-s\Lambda\bar{h}} \tag{2-22}$$

$$\phi_{fs} = (V_{rj} - V_{rb})\Phi_{fs} = (2V_{rj} - 1)\Phi_{fs} \tag{2-23}$$

where,

$$\Phi_{fs} = A_3(\Lambda\bar{h})^{\alpha_4} e^{-\alpha_5\Lambda\bar{h} + \alpha_6(\Lambda\bar{h})^2} \qquad \text{for } \Lambda\bar{h} \leq 7$$

$$\Phi_{fs} = 0 \qquad \text{for } \Lambda\bar{h} > 7$$

The values of $D, s, A_3, \alpha_4, \alpha_5, \alpha_6$ are listed in Table2-3 and 2-4.

Table 2-3. the values of D, s

γ	D	s	Range
1/9	1.51	0.52	$\Lambda\bar{h} > 1$
1/6	1.51	0.54	$\Lambda\bar{h} > 1$
1/3	1.47	0.58	$\Lambda\bar{h} > 1$
1	1.40	0.66	$\Lambda\bar{h} > 0.75$
3	0.98	0.79	$\Lambda\bar{h} > 0.75$
6	0.97	0.91	$\Lambda\bar{h} > 0.75$
9	0.73	0.91	$\Lambda\bar{h} > 0.75$

Table 2-4. the values of $A_3, \alpha_4, \alpha_5, \alpha_6$

γ	A_3	α_4	α_5	α_6
1/9	14.1	2.45	2.30	0.10
1/6	13.4	2.42	2.30	0.10
1/3	12.3	2.32	2.30	0.10
1	11.1	2.31	2.38	0.11
3	9.8	2.25	2.80	0.18
6	10.1	2.25	2.90	0.18
9	8.7	2.15	2.97	0.18

2.1.2.2 Thermohydrodynamic lubrication analysis of the engine bearing system considering surface roughness effect[7,11]

To perform the THD analysis, it is required to solve a generalized

Reynolds equation combining a 2-dimentional average flow model, energy equation, heat conduction equation and load equilibrium equation, which will be described in the following sections.

Generalized Reynolds equation based on the average flow model

Since the surface roughness effect and heat effect are simultaneously involved, a generalized Reynolds equation based on the average flow model is adopted in this book, which was deduced by Dr. Wang Xiaoli.

$$\frac{\partial}{\partial x}\left(\phi_x F_{20}\frac{\partial p}{\partial x}\right)+\frac{\partial}{\partial y}\left(\phi_y F_{20}\frac{\partial p}{\partial y}\right)=\frac{\partial}{\partial x}\left[U_h h_T +(U_0 - U_h)\frac{F_1}{F_0}\right]+\frac{U_0-U_h}{2}\sigma\frac{\partial\phi_s}{\partial x}+\frac{\partial h_T}{\partial t}$$

(2-24)

where, $F_{00}=\int_0^h \frac{dz}{\eta(z)}$, $F_{10}=\int_0^h \frac{zdz}{\eta(z)}$, $F_{20}=\int_0^h \frac{dz}{\eta(z)}\left(z-\frac{F_{10}}{F_{00}}\right)dz$,

$$F_0=\int_0^{h_T}\frac{dz}{\eta(z)}, \qquad F_1=\int_0^{h_T}\frac{zdz}{\eta(z)}$$

$$h_T=\frac{h}{2}\left[1+erf\left(\frac{h}{\sqrt{2}\sigma}\right)\right]+\frac{\sigma}{\sqrt{2\pi}}e^{-h^2/2\sigma^2},$$

$erf\,()$ is the error function.

A Schematic figure of the coordinate system of a bearing system is shown in Fig.2-1. For the crankshaft main bearing, the bush is stationary while the journal is rotational, thus $U_h = 0, U_0 = U$; The following dimensionless parameters are adopted,

$$\theta = x/R, \qquad \bar{y}=y/R, \qquad \bar{z}=z/h, \qquad \bar{h}=h/c=1+\varepsilon\cos\theta,$$

$$\bar{\eta}=\eta/\eta_0, \qquad \omega = U/R, \qquad \bar{p}=pc^2/\eta_0\omega R^2, \qquad \bar{t}=\omega t, \qquad \Lambda=c/\sigma$$

$$\bar{h}_T = 0.5\bar{h}\left[1+erf\left(\Lambda\bar{h}/\sqrt{2}\right)\right]+1/\left(\Lambda\sqrt{2\pi}\right)e^{-(\Lambda\bar{h})^2/2}$$

$$\bar{F}_1=\int_0^1\frac{\bar{z}}{\bar{\eta}}d\bar{z}, \qquad \bar{F}_0=\int_0^1\frac{1}{\bar{\eta}}d\bar{z}, \qquad \bar{F}_{20}=\int_0^1\frac{\bar{z}}{\bar{\eta}}\left(\bar{z}-\frac{\bar{F}_1}{\bar{F}_0}\right)d\bar{z}$$

Substituting them into equation (2-24), the dimensionless form is,

$$\frac{\partial}{\partial\theta}(\phi_x \, \overline{F}_{20}\overline{h}^3 \, \frac{\partial\overline{p}}{\partial\theta}) + \frac{\partial}{\partial y}(\phi_y \, \overline{F}_{20} \, \overline{h}^3 \, \frac{\partial\overline{p}}{\partial y}) = \frac{\partial}{\partial\theta}(\overline{h}_T \, \frac{\overline{F}_1}{F_0}) + \frac{1}{2}\sigma \, \frac{\partial\phi_s}{\partial\theta} + \frac{\partial\overline{h}_T}{\partial t}$$

$$(2\text{-}25)$$

The boundary conditions are,

at the oil inlet $\overline{p} = \overline{p}_0$, $\overline{p}_0$ is the pressure of inlet oil

at the outlet end, $\overline{p} = 0$

periodic boundary condition, $\overline{p}|_{\theta=0} = \overline{p}|_{\theta=2\pi}$

when the oil film breakage occurs, $\dfrac{\partial\overline{p}}{\partial\theta} = 0$ and $\overline{p} = 0$, that is the

Reynolds boundary condition.

Energy equation

If the body force and heat emission are neglected, energy equation of viscous fluid takes the following general form according to the principle of energy conservation,

$$\rho\frac{D(c_f T)}{Dt} = \nabla\cdot(K_f\nabla T) - \frac{T}{\rho}\frac{\partial\rho}{\partial T}\frac{Dp}{Dt} + \Phi \qquad (2\text{-}26)$$

where c_f is lubricant specific heat under constant pressure, K_f is heat conductive coefficient of lubricant. Φ is the dissipation work, for Newton liquid, it is,

$$\Phi = \eta\left[2\left(\frac{\partial u}{\partial x}\right)^2 + 2\left(\frac{\partial v}{\partial y}\right)^2 + 2\left(\frac{\partial w}{\partial z}\right)^2 + \left(\frac{\partial u}{\partial y} + \frac{\partial v}{\partial x}\right)^2 + \left(\frac{\partial u}{\partial z} + \frac{\partial w}{\partial x}\right)^2 + \left(\frac{\partial v}{\partial z} + \frac{\partial w}{\partial y}\right)^2\right]$$

$$(2\text{-}27)$$

When equation (2-26) is applied to unsteady hydrodynamic lubrication film, the following simplifications are assumed,

(1) The liquid is impressible.

(2) The constant-pressure specific heat c_f and heat conduction coefficient of lubricant K_f are constant.

(3) The film dimension in z direction is generally two orders of magnitude less than that in x and y directions, so the heat conduction along x and y directions are negligible.

(4) Similarly, comparing to $\dfrac{\partial u}{\partial z}$ and $\dfrac{\partial v}{\partial z}$, velocity gradients on other directions are also negligible.

Then, the energy equation can be simplified as,

$$\rho c_f \left(\frac{\partial T}{\partial t} + u\frac{\partial T}{\partial x} + v\frac{\partial T}{\partial y} + w\frac{\partial T}{\partial z} \right) = K_f \frac{\partial^2 T}{\partial z^2} + \eta \left[\left(\frac{\partial u}{\partial z}\right)^2 + \left(\frac{\partial v}{\partial z}\right)^2 \right] \qquad (2\text{-}28)$$

In addition to the dimensionless parameters introduced in equation(2-25), several other parameters are adopted as follows,

$$\omega = U/R, \bar{t} = t/\omega, \bar{u} = u/U, \bar{v} = v/U, \bar{T} = T/T_0, \rho = \rho/\rho_0$$

Referring to the approach in reference [15], a dimensionless energy equation derived from (2-28) is,

$$P_e \left(\bar{\rho}\frac{\partial \bar{T}}{\partial \bar{t}} \right) + \overline{\rho U}\frac{\partial \bar{T}}{\partial \vartheta} + \overline{\rho V}\frac{\partial \bar{T}}{\partial \bar{y}} + \bar{\Gamma}\frac{\partial \bar{T}}{\partial z} = \frac{1}{\bar{h}^2}\frac{\partial^2 \bar{T}}{\partial^2 \bar{z}^2} + \bar{\Phi} \qquad (2\text{-}29)$$

where, $P_e = \dfrac{\rho_0 c_f U}{RK_f}, \alpha = \dfrac{\eta_0 U^2}{K_f T_0}, \bar{\Phi} = \alpha \cdot \dfrac{\bar{\eta}}{\bar{h}^2}\left[\left(\dfrac{\partial \bar{u}}{\partial \bar{z}}\right)^2 + \left(\dfrac{\partial \bar{v}}{\partial \bar{z}}\right)^2 \right],$

$$\bar{\Gamma} = -\frac{1}{\bar{h}}\left[\frac{\partial}{\partial \bar{t}}\left(\bar{h}\int\bar{\rho}d\bar{z} \right) + \frac{\partial}{\partial\theta}\left(\bar{h}\int\overline{\rho u}d\bar{z} \right) + \frac{\partial}{\partial\bar{y}}\left(\bar{h}\int\overline{\rho v}d\bar{z} \right) \right],$$

T_0 is the inlet oil temperature, and ρ_0 is the density of lubricant under temperature T_0.

To determine boundary conditions of the film energy equation, the following two questions should be dealt with,
(1) Mixing effect of oil in the inlet groove
     During the working process of a journal bearing, oil film generates heat under viscous shear and results in an increase in temperature.

Then a part of the heated oil leaks from the ends of bearing, the other part returns to the inlet groove. Meanwhile, to keep the oil film complete and the bearing system cool, the engine lubricating system continuously supplies cool lubricant to the frictional pair. Therefore, the mixing effect of the heated oil and the cool oil should be considered to determine the actual oil temperature in inlet groove. If uniform distribution in the oil groove is assumed, the mixing oil temperature $\overline{T}_{mix}$ is[16],

$$\overline{T}_{mix} = \frac{Q_r}{Q_i + Q_r}\overline{T}_r + \frac{Q_i}{Q_i + Q_r}\overline{T}_i \tag{2-30}$$

where Q_r, Q_i are the recycled oil flux and the supplied cool oil flux respectively, T_r, T_i are the recycled oil temperature and the supplied cool oil temperature respectively.

(2) Gas cavity effect of oil film

For a journal bearing, the existence of radial convergence and divergence of film clearance may cause plus-minus variations in oil film pressure. Once a minus oil pressure exists, the oil film will be discontinuous, the resulting bubbles and strip film will cause gas cavities, as shown in Fig. 2-2.

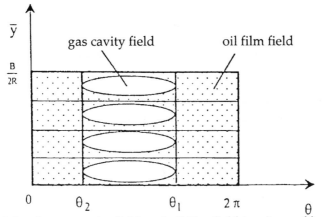

Fig. 2-2   the gas cavity field and oil film field in a journal bearing

The gas cavity effect plays a non-negligible role in the film pressure distribution as well as the lubrication performances of the whole bearing system. However, the energy equation mentioned above is generally suitable to a complete film field. How to apply it to the gas cavity field is a question that still needs to be solved. In the following section, an equivalent heat conduction coefficient $K(\theta)$ is introduced to represent the gas cavity effect, which is defined as[17],

in the oil film field, $\quad K(\theta) = K_f$

in the gas cavity field, $\quad K(\theta) = K_a - \dfrac{B(\theta)}{B}\left(K_a - K_f\right)$

Where K_f, K_a are the heat conduction coefficients of oil film and air respectively, B is the width of bearing, $B(\theta)$ is the effective film width at the gas cavity field. Since the oil pressure is zero at the gas cavity field, only the shear flow of film is considered. Based on the requirement of flow continuity, there is,

$$\frac{B(\theta)}{B} = \frac{h(\theta_2)}{h(\theta)} = \frac{\overline{h}(\theta_2)}{\overline{h}(\theta)}$$

The general boundary conditions for energy equation (2-29) are,

(1) for inlet slot, $\quad \overline{T} = \overline{T}_{mix}$ (2-31)

(2) symmetrical boundary condition, $\quad \left.\dfrac{\partial \overline{T}}{\partial y}\right|_{\overline{y}} = 0$ (2-32)

(3) on the interface between oil and journal,

$$\overline{T}_{i,j,0} = \overline{T}_s \quad \text{and} \quad \overline{q} = \int_0^{2\pi} \frac{1}{\overline{h}} \left.\frac{d\overline{T}}{d\overline{z}}\right|_{\overline{z}} d\theta = 0 \qquad (2\text{-}33)$$

It means that, the journal is an isothermal body, and the total thermal flux through the oil-shaft interface is zero. Dimensionless shaft temperature $\overline{T}_s$ is

$$\overline{T}_s = T_s / T_0 \qquad (2\text{-}34)$$

(4) on the interface between oil and bush

$$\left.\frac{\partial \overline{T}}{\partial \overline{r}}\right|_{\overline{r}=1} = -\frac{K(\theta)}{K_b}\frac{R}{C}\frac{1}{\overline{h}}\left.\frac{\partial \overline{T}}{\partial \overline{z}}\right|_{\overline{z}=1} \tag{2-35}$$

where K_b is the heat conduction coefficient of the bush, and Eq.(2-34) means that the thermal flux at the oil-bush interface is continuous.

(5) on the interface among oil, journal and bush, the THD boundary condition in Reference[18] is adopted.

Heat conduction equation

At the interface between lubrication film and solid, there exists a heat exchange. The interfacial boundary condition for energy equation is determined by the heat conduct ability of the contacting solids. The heat conduct equation of a solid takes the general form as,

$$\frac{DT}{Dt} = \frac{K}{\rho c}\nabla \cdot \nabla T \tag{2-36}$$

Taking the assumptions of (i) homogeneous and isotropic solid, (ii) constant heat conduct coefficient, specific heat and density, (iii) non-thermal-resource system, the heat conduction equation expressed in cylindrical coordinates takes the form as,

$$\frac{\rho c}{K}\frac{\partial T}{\partial t} = \frac{\partial^2 T}{\partial r^2} + \frac{1}{r}\frac{\partial T}{\partial r} + \frac{1}{r}\frac{\partial^2 T}{\partial \theta^2} + \frac{\partial^2 T}{\partial y^2} \tag{2-37}$$

This is the heat conduction equation applied to a bearing bush.

In addition to the dimensionless parameters introduced in equations (2-25) and (2-29), a new one is added, i.e., $\overline{r} = r/R$, then the dimensionless heat conduction equation is

$$\frac{\partial \overline{T}}{\partial \overline{t}} = \overline{a}\left(\frac{\partial^2 \overline{T}}{\partial \overline{r}^2} + \frac{1}{\overline{r}}\frac{\partial \overline{T}}{\partial \overline{r}} + \frac{1}{\overline{r}^2}\frac{\partial^2 \overline{T}}{\partial \theta^2} + \frac{\partial^2 \overline{T}}{\partial y^2}\right) \tag{2-38}$$

Where $\overline{a} = \dfrac{K_b}{c_b \rho_b R^2 \omega}$, $\overline{T}, c_b, K_b, \rho_b$ are the temperature, specific heat,

heat conduction coefficient and density of bush respectively.
The boundary conditions of equation (2-38) are,

(1) periodic boundary condition, $\left.\overline{T}\right|_{\theta=0} = \left.T\right|_{\theta=2\pi}$ (2-39)

(2) symmetric boundary conditions, $\left.\dfrac{\partial \overline{T}}{\partial \overline{y}}\right|_{\overline{y}=0}$ (2-40)

(3) on the interface between oil and bush, the boundary condition is the
     same as equation (2-33).

(4) on the interface between environment and bush,

$$\left.\frac{\partial \overline{T}}{\partial \overline{r}}\right|_{\overline{r}=\overline{R}_b} = -N_u\left(\left.\overline{T}\right|_{\overline{r}=\overline{R}_b} - \overline{T}_a\right) \tag{2-41}$$

Where, $\overline{T}_a = T_a / T_0, \overline{R}_b = R_b / R, N_u = \dfrac{h_b R}{K_b}$. h_b, K_b are heat convection

coefficient and heat conduction coefficient of bush. T_a is the environment

temperature and R_b is the external radius of bush.

(5) at the bearing end, $\left.\dfrac{\partial \overline{T}}{\partial \overline{y}}\right|_{\overline{y}=\pm\frac{B}{2R}} = -N_u\left(\left.\overline{T}\right|_{\overline{y}=\pm\frac{B}{2R}} - \overline{T}_a\right)$ (2-42)

Load equilibrium equations (referring to Fig.2-1)

When the film inertia is neglected, the motion of the journal centre
submits to the Newton's second law,

$$F_x + P_x = M_j \frac{dV_x}{dt}$$
$$\tag{2-43}$$
$$F_z + P_z = M_j \frac{dV_z}{dt}$$

where M_j is the journal mass, F_x, F_z are the load components in x and z

direction respectively. V_x, V_z are the movement velocities of journal centre

in x and z direction respectively. P_x, P_z are the countering forces of oil film in x and z direction respectively. They are listed as follows,

$$F_x = F \sin \gamma, \ F_z = F \cos \gamma \qquad (2\text{-}44)$$

$$P_x = P \sin(\delta + \varphi), \ P_z = P \cos(\delta + \phi) \qquad (2\text{-}45)$$

$$V_x = V \sin(\delta + \varphi), V_z = V \cos(\delta + \varphi) \qquad (2\text{-}46)$$

2.1.3 Criteria of acceptable design

Criteria for prevention of scratch

In order to prevent the bearing from scratching under a hydrodynamic lubrication, it is required that the minimum oil film thickness h_{min} be greater than an allowable value $[h_{min}]$. However, due to the great differences in structure and working condition of various engines, it is impossible to specify a unified allowable minimum film thickness. In practice, manufacturers of internal combustion engine determine the allowable value $[h_{min}]$ largely depending on their experience. In this book, referring to the literature on engine bearing design and Dr. Wang Xiaoli's THD analysis, two acceptable criteria are presented.

(1) Criterion of bearing minimum film thickness

$$h_{min} \geq [h_{min}] = 2(R_{zj} + R_{zb}) \qquad (2\text{-}47)$$

where R_{zj} and R_{zb} are the average surface roughness of the journal and the bush, respectively.

(2) Criterion of the maximum bush working temperature T_{max}

$$T_{max} < [T_{max}] \qquad (2\text{-}48)$$

where, the maximum allowable working temperature is determined by the bush materials. Table 2-5 lists the maximum allowable temperature of bush materials and other useful data from a large-scale engine manufacturer of China.

Criteria for frictional power loss

Decreasing frictional power loss is one of the main aims of engine tribological design. According to the expert experience collected by the authors, mechanical efficiency of a modern engine is $\eta_m = 0.70 \sim 0.80$ for a non-supercharged diesel engine, and $\eta_m = 0.80 \sim 0.90$ for a supercharged diesel engine. The frictional power loss can be up to 70% of the mechanical loss of an engine, in which frictional power loss from bearing system averagely accounts for 22%. Apparently, decreasing frictional power loss should be one of the aims of tribological design of an engine bearing system, which may be implemented by the lubrication analysis as well as the stand test.

Table 2-5. Service performance of bearing alloy

Materials	Mark	Maximum allowable pressure (MPa)	Allowable circular velocity (m/s)	Maximum working temperature(°C)	Journal Lowest hardness (HB)	Notes
Tin-base babbit metal	ChSnSb11-68-4	13	15 13	120 130	150	Light- load main bearing
Al-base babbit metal	ChPbSb 6-6		15	130		
Copper-lead alloy	ZQPb30 (2 layers)	25	8-10	170	300	Diesel engine
	ZQPb30 (3 layers)	24	10-13	150	230	
High-tin aluminium alloy	20% tin	30	13-15	170	230	
	30% tin	28		160	200	
6% low-tin aluminium alloy	2 layers	30	13-15	170	230	
	3 layers	35	15	170	230	Reinforced engine

2.1.4   Adjustable design parameters for tribological design

Adjustable design parameters of bearing system

Generally, adjustable design parameters involved in tribological design of bearing system are,
(1) surface topography of journal and bush, including surface direction parameter γ, surface roughness Λ and variance ratio between journal and bush V_{ij}.
(2) relative clearance between journal and bush ψ
(3) width-diameter ratio of bush λ
(4) different selections of bush width and diameter under the same λ
(5) location and length of oil groove

Adjustable design parameters of the entire set of engine

With the permission of the engine designer and the constraints of keeping the nominal power of an engine unchanged, the following collective parameters are adjustable in order to improve the lubrication properties, friction and wear performances of a bearing system [11],i.e.,

(1) Selection of lubricant, which determines the dynamic viscosity, viscosity-temperature property and viscosity-shear rate of lubricant.
(2) Bore-stroke ratio of engine. Since the piston swept volume of an engine is $V_h = \pi D^2 S$, different assemblies of piston stroke S and bore D exist to keep the V_h unchanged. According to the expert experience, there exists an appropriate selectable range for the engine bore-stroke ratio S/D under the situation that the working process of engine is not influenced.
(3) Piston swept volume and number of cylinders. The engine nominal power is,

$$N_e = \frac{P_e \times V_h \times i \times n}{300\tau} (kW) \tag{2-49}$$

where p_e — cylinder average pressure in a work circle of engine

V_h — piston swept volume of a cylinder

i — number of the cylinders

n — rotational speed of crankshaft (r / min)

τ —   stroke number

Under the precondition of keeping the engine nominal power and the product of V_h and i unchanged, different assemblies of V_h and i can be designed.

(4) Average cylinder pressure and rotational speed. According to the same precondition mentioned above, keeping the nominal power and the product of p_e and n unchanged, different assemblies of p_e and n can be designed.

(5) Mass of the reciprocal moving component. In addition to decreasing the rotational speed, decreasing the mass of a reciprocal moving component will also result in a decreasing inertia force, which is one of the most important approaches to improve the lubrication condition and frictional power loss of a bearing system.

2.2   Tribological design for the piston assembly of an engine

2.2.1 General consideration of design principles

Piston assembly of an engine includes three frictional pairs, i.e., piston ring-cylinder liner pair, piston skirt-cylinder liner pair and piston ring-ring slot pair. The former two pairs are considered in this book.

The aims of tribological design of a piston assembly include two aspects:
(1) Wear control. Wear life of each frictional pair is required to be longer than a predefined operating requirement. As a scratch problem in a

piston assembly is not the same important as that in a bearing system which always leads to disastrous results, therefore, prevention of the scratch is not treated as a essential design aim of a piston assembly, which is just ensured via wear control instead.

(2) Decreasing frictional power loss. It has an important role to play in the tribological design of piston assemblies.

Since R.A. Castleman[19] first performed lubrication analysis on the piston ring-cylinder liner friction pair based on the hydrodynamic lubrication theory, much effort has been made to investigate the friction and wear performances of piston assembly. Consequently, tribological design of piston assembly may be established and performed. Modern research on the lubrication analysis of a piston system is summarized as follows.

(1) Lubrication analysis considering the squeeze effect. Based on Reynolds equation, Castleman[19] only considered the dynamic effect in his lubrication analysis of piston ring-cylinder pair, and therefore, suggested that an arc shape should be adopted as the radial profile of a piston ring face. According to his analysis, oil film thickness of the ring at the top and bottom dead centres was zero, which resulted in an enlarged frictional power loss at or near the two transient points. S.Furunhama[20] and T.Lloyd[21] added the film squeeze effect into their lubrication analysis and verified their results successfully by experiments. The results indicated that although the oil film thickness at the top and bottom dead centres was very small, it was not zero.

(2) Lubrication analysis considering the insufficient oil supply. In 1979, D. Dowson et al[22,23] expanded their lubrication analysis of a single piston ring to the piston ring packs and considered the interaction of oil flow among the piston rings. Obviously, the existence of the front ring will reduce the oil supply amount for the following ring, which will result in insufficient oil supply problems. In order to keep the oil flow continuous in the piston ring pack, they amended the allowable

oil film thickness for a sufficient supply.

(3) Lubrication analysis considering the surface roughness.Up to the 80s of last century, lubrication analysis of piston ring-cylinder liner pair had been built on smooth surfaces. The lubrication state was considered as either hydrodynamic lubrication or boundary lubrication. After the average Reynolds equation based on rough surfaces was proposed by Patir and Cheng in 1979, combining this model with the Greenwood and Tripp's asperity contact model[24], S.M. Phode et al[25] performed their prominent work on the lubrication analysis of the piston ring–cylinder liner. It was theoretically demonstrated that, both the mixed lubrication state (near the top and bottom dead centres) and the hydrodynamic lubrication state (at the middle point) existed during the whole stroke of a piston, and the friction forces were maximum at the dead points. The contacting load acting at the contacting asperities under the mixed lubricant state could also be calculated.

(4) Lubrication analysis considering the secondary motion of piston skirt. Research on lubrication analysis of piston skirt-cylinder liner pairs began much later than that of the ring-cylinder liner. G.D. Knoll[26] investigated the lubrication state of a piston skirt in 1982, and calculated the hydrodynamic loading capacity. However, the results were not in a scientific sense since the secondary motion of piston skirt was neglected. Coupling the dynamic equation with the hydrodynamic lubrication equation, D.F. Li et al[27] proposed a perfect lubrication equation of a piston skirt in 1983. During their solution of Reynolds equation, the squeeze effect and hydrodynamic effect were treated separately and then added together, however the surface roughness effect was also neglected. In 1991, D. Zhu et al[28] solved these problems successfully, which greatly promoted the lubrication analysis of piston skirt.

(5) Lubrication analysis considering the non-uniform distribution of circumferential spring forces of piston rings. Considering the non-uniform distribution of circumferential spring forces of piston rings, Y.Z. Hu et al[29] proposed a lubrication model to describe the non-uniform distribution features of circumferential film of piston rings. K.Liu et al[11,30,31,39] performed a 2-dimensional lubrication analysis on a piston ring pack additionally considering the secondary motion of piston skirt.

(6) Lubrication analysis considering the comprehensive effects mentioned above. K.Liu et al[30-32] analysed the lubrication performances of piston ring-cylinder liner pairs and piston skirt-cylinder liner pairs considering the comprehensive effects including the film squeeze effect, hydrodynamic effect, insufficient oil supply, surface roughness, the secondary motion of a piston skirt, and the non-uniform distribution of circumferential ring spring forces. According to the expert experience collected in this book, this analysis method is available for the low- reinforced internal combustion engine.

(7) Lubrication analysis considering the temperature field of piston assembly. C.L. Gui[11] , one of the authors, proposed that, for a moderately or highly reinforced engine, it is necessary to consider the temperature field in lubrication analysis of piston assembly, and it is also important to determine the viscosity-temperature property of the lubricant based on the inwall temperature of the cylinder liner.

As to the research on friction force of piston assembly, since the surface roughness effect is involved, the contact behaviour on rough surfaces can be described according to Greenwood and Tripp's asperity contact model[25]. Furthermore, the friction force resulting from the asperity contact can be obtained, which establishes the base for wear predication. Furuhama et al[20,33-35] have performed a profound, continuous and particular research on friction mechanism of piston ring-cylinder liner pairs. Benefiting from their developed experimental equipment which can overcome the inertia effect, they measured the variations of friction force during the whole stroke of a piston. It indicated that the friction force of a piston ring comprised of two parts, the hydrodynamic friction force due to mixed lubrication and the solid friction force due to local asperity contact.

In addition, wear mechanism of piston assembly is very complicated. The influencing factors include lubrication state, heat, external or self-produced abrasive, erosion, load, etc. So it is nearly impossible to establish a precise model including all the mentioned factors to describe wear properties of a piston assembly. Therefore, it is more practical to adopt a simple model to include the common and main influencing factors. One of the representative models is Archard model[36], L.L. Ting[37] applied it into wear predication of piston ring-cylinder liner pairs. C.L. Gui[11,38] adopted it in wear analysis of a general mechanical component. C.L. Gui's method are adopted in this book.

2.2.2   Mathematic equations for deign and calculation

2.2.2.1   Comprehensive lubrication analysis of piston assembly[11,30-32,39]

To involve the comprehensive factors including film squeeze effect, insufficient oil supply, surface roughness, secondary motion of a piston skirt, non-uniform distribution of circumferential ring spring forces into lubrication analysis of piston assembly, it is required to solve simultaneous equations including the dynamic equation of piston, 2-dimensional average Reynolds equation, film thickness equation, load equation, flow equation of insufficient oil supply and the asperity contact equation. Therefore, the axial and circumferential oil film thickness for piston ring-cylinder liner pair and skirt-cylinder liner pair, as well as the variations of friction power loss in a working period of an engine can be calculated. Besides, whether the offset of piston skirt will result in knocking on cylinder wall or not can be judged theoretically.

Piston dynamic equation(referring to.2-3)

$$\begin{bmatrix} m_{pin}\left(1-\dfrac{a}{L}\right)+m_{pis}\left(1-\dfrac{b}{L}\right) & m_{pin}\dfrac{a}{L}+m_{pis}\dfrac{b}{L} \\ \dfrac{I_{pis}}{L}+m_{pis}(a-b)\left(1-\dfrac{b}{L}\right) & m_{pis}(a-b)\dfrac{b}{L}-\dfrac{I_{pis}}{L} \end{bmatrix}\begin{bmatrix} \ddot{e}_t \\ \ddot{e}_b \end{bmatrix}=\begin{bmatrix} F+F_s+F_f tg\phi \\ M+M_s+M_f \end{bmatrix} (2\text{-}50)$$

where m_{pin}, m_{pis} — masses of piston pin and piston

$\qquad I_{pis}$ — inertia moment of piston

$\qquad e_t, e_b$ — eccentricities of the top and bottom of piston skirt

$\qquad a, b$ — distances from the top end of the piston to the centre of the piston pin and the mass centre of the piston.

$\qquad L$ — length of piston skirt

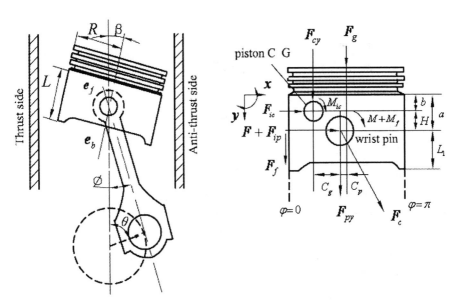

Fig. 2-3   Geometric dimensions and the secondary movement of piston skirt

F, M —the force and the moment resulting from hydrodynamic motion between the piston skirt and cylinder liner.

F_f, M_f —friction force and friction moment resulting from the reciprocal movement of the piston skirt.

$$F_s = \left(F_G + F_{py} + F_{cy} \right) \mathrm{tg}\phi, \qquad M_s = F_G C_p - F_{cy} C_G$$

F_G — force acting on the piston resulting from the gas pressure of the combustion chamber

F_{py}, F_{cy} —inertia forces resulting from the reciprocal movement of the piston pin and piston

ϕ —included angle between connecting rod and the cylinder axis

C_p —wrist pin offset

C_G —radial distance between the centre mass of piston and the centre of wrist pin.

So the trajectory of the piston skirt's secondary motion may be obtained by solving the above equations.

The relationship between the piston's offset angle and the transverse displacement of the piston pin, and the top and bottom of the skirt is approximately described as,

$$e_t = e_p + \alpha\beta$$
$$e_b = e_p - (L - \alpha)\beta \tag{2-51}$$

Their dimensionless form is introduced and represented as

$$E_t = \frac{e_t}{C_t} \qquad E_b = \frac{e_b}{C_b} \tag{2-50'}$$

where C_t, C_b are the radial clearances at the top and bottom ends of the piston. Obviously, when $|E_t| > 1,$ or $|E_b| > 1$ exists, a piston knock may occur.

2-dimensional average Reynolds equation

$$\frac{\partial}{\partial x}(\phi_x \frac{h^3}{12\eta}\frac{\partial p}{\partial x}) + \frac{\partial}{\partial y}(\phi_y \frac{h^3}{12\mu}\frac{\partial p}{\partial y}) = \frac{u}{2}\frac{\partial \overline{h}_T}{\partial x} + \frac{u}{2}\sigma\frac{\partial \phi_s}{\partial x} + \frac{\partial \overline{h}_T}{\partial t} \qquad (2\text{-}53)$$

where p —average hydrodynamic pressure

η —dynamic viscosity of lubricant

u —piston velocity

$\overline{h}_T$ —average film thickness

ϕ_x, ϕ_y —pressure flow factors, referring to equation(2-19)

ϕ_s —shear flow factors, referring to equation(2-20)

σ —composite surface roughness

t —time

Oil film thickness equation

(1) For the piston ring-cylinder liner pair. Considering the piston offset due to the secondary motion, film thickness equation is,

$$h = h_0 + \frac{4(l_b \sin\beta + e_0 \cos\beta)e_0 \cos^4\beta}{l_b^2(2e_0 \cos^2\beta + l_b \sin\beta)^2}y^2 \qquad (2\text{-}54)$$

where e_0 is the bulging height when a symmetric parabolic curve is adopted as the profile line of the piston ring face, as shown in Fig.2-4.

(2) For skirt-cylinder liner pair. Considering the piston offset due to the secondary motion, film thickness equation is expressed as,

$$h = c + e_t \cos\varphi + (e_b - e_t)\frac{Y}{L}\cos\varphi + f(Y) \qquad (2\text{-}55)$$

where c —radial clearance between the skirt and the cylinder liner

φ —circumferential angle around the piston axis

Y —longitudinal coordinates from the top end of the skirt

$f(Y)$ —profile function of the piston skirt

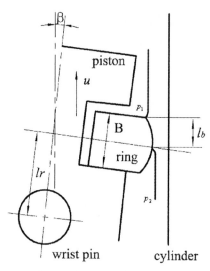

Fig. 2-4   Lubrication analysis of ring-cylinder liner considering offset of piston

Load equation for piston ring

The stress and deformation of the piston ring play a significant role in the film thickness of ring-cylinder liner pair. In order to determine the variations of film thickness along the circumferential direction of the ring, the ring is circumferentially divided into m elements (referring to Fig.2-5). The load acting on each element includes five items.

(1) Cylinder pressure acting on the back of the piston ring, p_1, which is obtained from load indicator diagram of engine.

(2) Spring force of the piston ring, p_e, according to Reference[32]or[41]

$$P_e(\varphi) = p_0 \left\{ n + l\left[\frac{1}{2}(\varphi^2 + \sin^2 \varphi) + 2(\cos\varphi + \cos^2 \varphi)\right]\right\} \qquad (2\text{-}56)$$

φ — circumferential angle from the gap of the piston ring

p_0 — pressure constant

n, l — contribution property coefficients of contact pressure, there is,

$$n + 2.894l = 1$$

After the n is randomly selected, l may be determined, and then the pressure distribution properties are known.

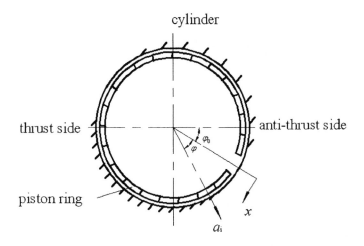

Fig. 2-5　circumferential coordinate system of a piston ring

(3) Inertia force due to the accelerated velocity of the piston's secondary motion, Δma_i .

Δm —mass of a ring element.

a_i —local acceleration of the ring element, it is related to the local transverse acceleration,

$$a_i = a_r (\cos \varphi_0 + \varphi) \qquad (2\text{-}57)$$

where a_r is $\qquad a_r = \ddot{e}_p + l_r \ddot{\beta}$

$\ddot{e}_p$ —offset acceleration of the centre of piston pin

$\ddot{\beta}$ —transverse acceleration of the centre of the piston pin

l_r —axial distance from the ring position to the centre of the piston pin, as referring to Fig.2-4.

(4) Oil film pressure p , which is determined by Reynolds equation.

(5) Asperity contact pressure p_A, which is determined from the asperity contact equation.

Therefore the composite force acting on a ring element is

$$w_i\left(h_i, \dot{h}_i\right) = \iint p_1 dy dx + \iint p_e dy dx + \Delta m a_i - \iint (p + p_A) dy dx \qquad (2\text{-}58)$$

According to D.C. Sun[40], if oil film support exists between a ring and a cylinder liner, there is,

$$w_i\left(h_i, \dot{h}_i\right) = 0 \qquad (2\text{-}59)$$

For the whole piston ring, there is,

$$W\left(H_i, \dot{H}_i\right) = 0 \qquad (2\text{-}60)$$

where

$$W = \left[w_1, w_2 \cdots, w_m\right]^T,$$
$$H = \left[h_1, h_2 \cdots, h_m\right]^T,$$
$$\dot{H} = \left[\dot{h}_1, \dot{h}_2 \cdots \dot{h}_m\right]^T$$

Consideration of starved lubrication of the piston rings

For a piston ring pack, the insufficient supply of lubricant makes it necessary to consider the starved lubrication. As shown in Fig.2-6, lubricant supply condition of the following ring is controlled by that of the front ring. If the circumferential flow of lubricant is negligible, and the output flux of the front ring is $q_{i,out}$, the input flux of the following ring is $q_{i+1,in}$. It is obvious that, when $q_{i,out} > q_{i+1,in}$ exists, the following ring operates in a sufficient lubrication regime, while if $q_{i,out} < q_{i+1,in}$ exists, the following ring is in a starved lubrication regime. Therefore, lubricant flow equation between piston rings is

$$q_{i+1,in} = q_{i,out} \tag{2-61}$$

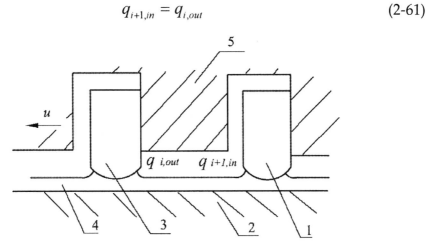

Fig. 2-6   Models of starved lubrication of piston rings.
1- Ring $i+1$, 2- Cylinder, 3- Ring i, 4- Lubricant, 5- Piston.

The lubricant flux is related to the pressure gradient and the film thickness as follows,

$$q = -\phi_y \frac{h^3}{12\eta} \frac{\partial P}{\partial y} + \frac{u}{2} \bar{h}_T + \frac{u}{2} \sigma \phi_s \tag{2-62}$$

where $\dfrac{u}{2}\sigma\phi_s$ represents the additional flow resulting from the sliding movement of rough surfaces.

The starved lubrication equation of a piston ring pack is obtained by combining equation (2-60) and (2-61), which is non-linear.

$$\begin{cases} W\left(H_i, \dot{H}_i\right) = 0 \\ q_{i+1,in} = q_{i,out} \end{cases} \tag{2-63}$$

Asperity contact equation

When the film thickness between two rough surfaces is less than a certain value, the friction pair operates in a mixed lubrication regime, asperity contact exists only at a part of the interface. Load acting on the ring-cylinder liner interface is carried by the lubricant as well as the contacting asperities. According to Greenwood and Tripp[25], under an assumption of elastic deformation, loading capacity of the asperities is,

$$W_A = \left(\frac{16\sqrt{2}}{15}\right)\pi\left(D_{sum}\rho\sigma\right)^2 E'\sqrt{\frac{\sigma}{\rho}}AF_{\frac{5}{2}}\left(\frac{h}{\sigma}\right) \tag{2-64}$$

The actual contact area is

$$A_c = \pi^2\left(D_{sum}\rho\sigma\right)^2 AF_2\left(\frac{h}{\sigma}\right) \tag{2-65}$$

D_{sum} — number of the summits per unit area

ρ — average curvature radius of the summits

A — nominal contact area

E' — equivalent elastic modules

$F_n\left(\frac{h}{\sigma}\right)$ — function of the ratio of film thickness to surface roughness

Therefore, the pressure acting on the asperity contact area is

$$p_A = \frac{W_A}{A_c} \tag{2-66}$$

Shear resisting force acting at the sliding surfaces is

$$\tau_A = \tau_0 + fp_A \tag{2-67}$$

τ_0 — shear strength of the lubrication film

f — friction coefficient

Frictional power loss

Friction forces between the piston ring-cylinder liner pair and skirt-cylinder liner pair are both composed of two parts, i.e., shear resisting force and asperity contact friction force.

$$F = \int_A \left(\tau_A + \tau_H \right) dA \qquad (2\text{-}68)$$

The hydrodynamic shear resisting force is also made up of two parts,

$$\tau_H = \tau_{H1} + \tau_{H2} \qquad (2\text{-}69)$$

$$\tau_{H1} = \frac{\eta u}{h} \left(\phi_f + \phi_{fs} \right) + \phi_{fp} \frac{h}{2} \frac{dp}{dx} \qquad (2\text{-}70)$$

$$\tau_{H2} = \left(\frac{\sigma_1}{\sigma_2} \right)^2 \left(\phi_{fp} h - \overline{h}_T \right) \frac{dp}{dx} - \frac{2\eta u}{h} \phi_{fs} \qquad (2\text{-}71)$$

where $\phi_f, \phi_{fs}, \phi_{fp}$ are the shear stress factors, which are determined by equations (2-21) to (2-23).

Therefore, the frictional power loss N_f is

$$N_f = F \times u \qquad (2\text{-}72)$$

2.2.2.2 Wear prediction method of piston ring assembly

Based on the Archard wear predication model[36]

$$R_v = \frac{V}{L} = k \frac{P}{H} \qquad (2\text{-}73)$$

It can be derived that[38] ,

$$K_c = \frac{VH}{NL} = \frac{VH}{Nvt} = \frac{(V/A)H}{(N/A)vt} = \frac{hH}{P_a vt} \qquad (2\text{-}74)$$

where R_v ─wear volume rate

V ─wear volume

L ─continuous were distance of friction pair

k ─wear coefficient

P ─load

H ─ surface hardness of the soft materials

K_c — calculated wear coefficient

v — sliding velocity

t — required working life (continuous friction time)

A — nominal contact area

h — allowable wear depth during the wear life

P_a ⌐nominal contact pressure derived from Eq. (2-66)

$$P_a = \frac{16\sqrt{2}}{15}\pi\left(D_{sum}\rho\sigma\right)^2 F_{\frac{5}{2}}\left(\frac{h}{\sigma}\right) \tag{2-75}$$

Let K_t be the experimental wear coefficient obtained from simulation test or engine stand test, in which all the components adopted were selected from the serving parts, and the lubricant, environment temperature, load and pressure were as the same as the real working conditions.

If $K_c < K_t$ (2-76)

It indicates that, the predicated wear life does not satisfy the demand, and the design should be modified.

If $K_c > K_t$ (2-77)

It means that the predicated wear life satisfies the demands.

2.2.3.   Criteria of acceptable design

Criteria of wear life

Generally, the first heavy repairing cycle of an engine is determined by the time when the cylinder liner reaches its limit dimensions. Table 2-6 is the maximum allowable wear limit of a cylinder liner based on the expert experience collected from China market by the authors. On the other hand, the cumulative operating hours of engines before their first heavy repairing circle are listed in Table 2-7.

Table2-6   *Maximum allowable wear limit of a cylinder*

Cylinder diameter D (mm)	50~100	100~200	200~400	400~800
Allowable relative wear limit	(1/500~ 1/200) D	$\dfrac{1}{800}D$	$\dfrac{1}{400}D$	$\dfrac{1}{200}D$

Table 2-7 the cumulative operating hours of an engine before its first heavy repairing circle

Engines	The cumulative operative hours
Farm diesel engine	6000~8000
Engine of Small high-speed vehicle	8000~16000 (about 150 thousand~200 thousand kilometers)
Truck engine	7500~15000 (about 100 thousand~150 thousand kilometers)
Industrial diesel engine	6000~15000
Loco engine	8000~32000
Marine diesel engine	15000~80000

According to the data listed above and equation (2-74), the predicated wear coefficient of a cylinder liner can be calculated. On the other hand, based on the data accumulated in their enterprise, tribology engineers may also calculate wear coefficient. The acceptable wear condition for a cylinder liner has been expressed in equation(2-77).

A piston ring is generally a renewable component, it also has the demands on wear-resistant life and wear limit, which are related to the materials and surface coatings adopted by each manufacture, therefore, each enterprise owns its relevant design data.

A piston is generally considered to have the same wear life as the entire set of engine, and neither heavy repair nor renewal is allowable. That is why those manufactures specify their strict limits on the piston skirt wear. For example, the skirt of a heavy marine diesel engine is usually treated with an antifriction coat in order to reduce the skirt wear to zero during the useful life period.

Criteria of frictional power loss

The average frictional power loss of a piston assembly occupies about 70% of the whole frictional power loss of an engine. Therefore, decreasing frictional power loss of piston assembly is especially emphasized in order to improve the mechanical efficiency. An engine tribology engineer should pay close attention to decreasing frictional power loss of the piston assembly.

Criteria of engine knock

It is indicated from equation (2-52) that, the prevention criteria of engine knocking resulted from the piston skirt's transverse movement is,

$$|E_t| < 1 \quad \text{and} \quad |E_b| < 1 \tag{2-78}$$

2.2.4 Adjustable design parameters of tribological design

Adjustable design parameters of piston assembly

(1) Piston skirt profile parameters. Piston skirt profile parameters include axial directional profile parameters and radial directional profile parameters. The axial directional profile plays a significant role in the offset angle of piston's secondary motion. Therefore, it influences the film thickness of the piston ring-cylinder liner pair and that of skirt-cylinder liner pair, and results in engine's knocking behavior. Radial directional profile is designed to deform into a circular shape when heated during the working process, so as to adapt to the cylindrical face of the liner. Therefore, radial directional profile is initially designed as an elliptic shape when the engine is idle, instead of a circular shape. Relevant expert experience and design data have

been collected in the intelligent system, which are optional during the tribological design process.

(2) Clearance between the skirt and cylinder liner. Since the piston skirt profile is non-cylindrical in axial direction and non-circular in radial direction, the nominal dimension at the middle or lower middle location of the piston skirt is designed to be equal to the nominal inner diameter of the cylinder. The fit clearance considering the mutual tolerance is named cylinder-fit clearance. Too large a cylinder-fit clearance will result in engine knocking, while too small a cylinder-fit clearance will cause scratching.

(3) Offset of piston pin. The offset location of a piston pin influences the secondary motion of the piston skirt, the resulting piston beat and transverse acceleration also influence the lubrication performances of the piston assembly.

(4) Length of piston skirt. The length of the piston skirt directly determines the loaded area between the skirt and the liner, and as a consequence, influences the secondary motion, film thickness, as well as the friction force between the skirt and the liner.

(5) Spring force of piston ring. The effects of the spring force of the piston ring involve two aspects: distribution properties of the spring force and the average spring force. The former one influences the uniformity of the circumferential film thickness, and the latter one influences the frictional power loss.
(6) Width of piston ring
(7) Profile shape and dimension of ring face
(8) Surface roughness

Adjustable design parameters of the whole set of engine
It is the same case as that in section 2.2.4

2.3 Tribological design for the valve train system of an engine

2.3.1 General consideration of design principles

The valve train system of an engine includes three friction pairs, i.e., cam-follower pair, valve-valve seat pair and valve-valve guide pair. The former two pairs are investigated in this book.

The main aim of tribological design for a valve train system is the prevention of wear, including pitting (slighter than normal wear failure) and serious scuffing. In a practical sense, the working lives of the three involved friction pairs are all determined by their wear-resisting properties. The wear region of the cam-follower pair is generally located at the cam nose. In the cases of a serious wear, it has been reported that the cam nose was ground off even several millimeters only after several hours or dozens of hours. The serious wear of a cam nose will result in a great decrease in valve lift range and rotational speed, as well an undesired increase in engine temperature and oil consumption, therefore will make the engine lose its normal operational ability.

A valve-valve seat pair works in an extremely austere lubrication state. If improperly designed, the valve-valve seat pair will suffer from serious wear after operating dozens of hours or even several hours. The fall-in distance of valve may even reach 1~2 mm or more, which will lead to seal failures.

Lubrication analysis of a cam-follower pair is closely related to the development of elastrohydrodynamic lubrication theory. Holland[41] made a significant achievement in the cam-follower lubrication analysis, considering the load and velocity between the cam and the followers as functions of time. Therefore, he proposed that between the cam and followers there exist not only tangential movement but also normal approach movement, i.e., the squeeze effect. Moreover, an unsteady elastrohydrodynamic lubrication analysis approach for the cam-follower pair was established. Based on the same ideas, Dowson[42] and Mei[43] developed a numerical method for quickly solving the elastrohydrodynamic equation in their following research. Especially, considering the surface roughness effect, Mei adopted the average Reynolds equation to calculate the variation of film thickness ratio, which interpreted the wear mechanisms of the cam-follower pair very successfully. In his research, besides the perfect elastrohydrodynamic lubrication theory, lubrication analysis of the cam-follower pair is also in agreement with the experimental results.

However, various presented calculations demonstrated that, the minimum film thickness at the cam nose is less than 0.5 μ m. Based on the Holland model, H.Q. Yu et al[44] analysed film thickness and the ratio of film thickness to composite roughness ratio with various cam profile lines of engines. Their results indicated that, the film thickness-roughness ratio is less than 1 at the minimum film region, corresponding with the boundary lubrication state, and the film thickness-roughness ratio is larger than 4 at the thick film region corresponding with the complete elastrohydrodynamic lubrication state. The variations of the film thickness-roughness ratio interpreted the real wear mechanism of the cam and followers.

From the above relations it can be concluded that, the lubrication state at the cam nose region cannot be improved via changing the design or modifying the elastrohydrodynamic analysis method. Therefore researchers turn to other approaches to perform cam tribological design. Basing on Dowson-Higginson's elastrohydrodynamic lubrication calculation of line contact as shown in Equation(2-79), Deschler and Wittmann[45] proposed a simplified equation to calculate the minimum film thickness as equation (2-80), in which the equivalent elastic module E' and viscosity-pressure coefficient α were considered as constant since the cam and tappet were usually made of steel and the lubricant was always mineral oil. Meanwhile, the load on unit contact length W/L plays a minor role in the minimum film thickness, so it is negligible.

$$h_{min} = \frac{1.6\alpha^{0.6}(\eta_0 U)^{0.7} R^{0.43} E'^{0.08}}{(W/L)^{0.13}} \qquad (2\text{-}79)$$

$$h_{min} = 1.6\times10^{-5}\sqrt{\eta_0 UR} \qquad (2\text{-}80)$$

According to the structural and geometrical features of the cam and tappet, minimum film thickness for a pair with a stationary tappet and a punch head is,

$$h_{min} = K_0(r_0 + s)\sqrt{\left|2\left(\frac{\rho}{r_0+s}\right)^2 - \frac{\rho}{r_0+s}\right|} \qquad (2\text{-}81)$$

where $K_0 = 1.6\times10^{-5}\sqrt{\dfrac{\omega\eta_0}{2}}$

$r_0, s, \rho -$ referring to Fig.2-7

$\omega -$ rotational speed of cam

$\eta_0 -$ dynamic viscosity of lubricant under constant pressure and temperature

Based on equation(2-81), Deschler and Wittmann proposed a dimensionless critical parameter to identify the lubrication state of a cam, namely critical hydrodynamic index Nr.

$$N_r = \frac{\rho}{r_0 + s} \qquad (2\text{-}82)$$

So there is

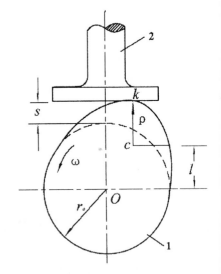

Fig.2-7 cam-tappet friction pair

$$h_{\min} = K_0 (r_0 + s) \sqrt{\left| 2N_r^2 - N_r \right|}$$

$$= K_0 \frac{\rho}{N_r} \sqrt{\left| 2N_r^2 - N_r \right|} \tag{2-83}$$

Obviously, when Nr=0 or Nr=0.5, $h_{\min}$ =0; when Nr is between 0 and 0.5, $h_{\min}$ has a maximum value, corresponding with an optimal lubrication state. The corresponding optimal film thickness is noted as h_{opt}, and the Nr is noted as N_{opt}.

Furthermore, a relative film thickness $\overline{h}_\xi$ is defined as

$$\overline{h}_\xi = \frac{h_{\min}}{h_{opt}}$$

Since $h_{opt} = K_0 (r_0 + s) \sqrt{\left| 2 \times 0.25^2 - 0.25 \right|} = 0.35355 K_0 (r_0 + s)$ (2-84)

there is $\quad \overline{h}_\xi = \dfrac{h_{\min}}{h_{opt}} = \dfrac{\left| (2N_r - 1) \right|^{0.5} N_r^{0.5}}{0.35355}$ (2-85)

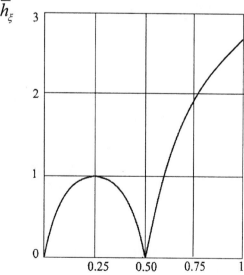

Fig. 2-8   the relationship between $\overline{h}_\xi$ and Nr

The relationship curve between $\overline{h}_\xi$ and Nr is illustrated in Fig.2-8. In general, a thin film thickness region exists for a stationary tappet with a punch head, which corresponds with Nr between 0 and 0.5. Moreover, when Nr is greater than 0.5, the minimum film thickness will increase with Nr. Therefore, in order to decrease the wear of cam, the working region should be selected with a larger Nr value, especially away from the region of $0<Nr<0.5$. However, the increasing of Nr will result in other problems. Due to the dilemma encountered in engineering, the Nr value at the cam nose region is still generally set in the range of $0\sim0.5$ for the friction pair with a stationary tappet and a punch head. But the designed Nr should be submitted to the equation $N_r \leq N_{ropt}$, as it will gradually approach to N_{ropt} during the wear process. On the contrary, if the designed N_r is larger than N_{ropt}, it will approach to 0.5 during working process, which indicates that the slight wear has deteriorated into a scratch. Therefore, N_r value at the cam nose region should be designed in the range of 0.15 to 0.25.

In addition, based on the concept of lubrication property index proposed by Benz, R. Müller[46] presented another lubrication property index of cam-tappet pair, noted as N_s. Furthermore, considering the lubrication state and stress, M. Ryti[47] investigated the optimal design of a cam profile, and proposed a new dimensionless lubrication property index. It has been demonstrated that[48], a definite relationship exists among the critical hydrodynamic index N_r, lubrication property index N_s and the new dimensionless lubrication property index. Therefore, either of the indexes is valid to describe the wear properties of a cam-tappet pair. The critical hydrodynamic index Nr is adopted in this book.

Besides the lubrication property assessment, contact pressure should also be analysed for an initial design in order to control the cam-tappet wear. During working process, a valve-valve seat pair also carries extremely heavy mechanical and thermal load, and endures the erosion of the corrosive gas. It is apt to be severely worn, which will result in the sinkage of a valve, increase in harmful combustion volume, deterioration of engine performance, and even cause seal failure of the valve-valve seat. Therefore, the basic aim of tribological design of a valve-valve seat pair is wear control. Correspondingly, the following calculations should be performed [11,49].

(1) Dynamic movement analysis. When the gas valve is first opened and then drops down to its close location, due to the combination of valve spring and moving part inertia, the valve seat will suffer a sudden impact, which will result in impact wear on the valve and the valve seat. The impact wear is related to the impact magnitude. Therefore, the valve dropping movement including dropping velocity and dropping region should be analysed in order to control the impact wear of valve.

(2) Stress of the valve head. When the valve is closed, a valve disk carries the gas pressure of the combustion chamber and yields a bending stress. The stress should be controlled not only to satisfy the demand on bending strength, but also to reduce indirect abrasion on the interface of the valve and valve seat.

(3) Deformation of the valve head. It has been demonstrated that, in addition to the impact wear, fretting wear also exists on the interface between the valve and valve seat, which are related to the elastic deformation of the valve. Therefore, the deformation of the valve head should be evaluated to control the wear performance.

2.3.2 Calculation equation

Cam-tappet friction pair
(1) Critical hydrodynamic index Nr[11,44,48], referring to Equation.(2-82).
(2) Calculation of contact stress[11]

$$\sigma_c = 0.418 \sqrt{\frac{PE_m}{B\rho_1}} \leq [\sigma_c] \tag{2-86}$$

where, P —contact load between cam and tappet

E_m —equivalent elastic module of cam and tappet

$$E_m = \frac{2E_1 E_2}{E_1 + E_2} \tag{2-87}$$

B —contact width of the cam-tappet pair

ρ_1 —curvature radius of cam at the contact point

$[\sigma_c]$ —allowable contact stress

Valve-valve seat friction pair

(1) Dynamic movement analysis of valve[11,49](referring to Fig.2-9)

$$m\frac{\partial y^2}{\partial t^2} = K_x \left(y_c - y - x_0 \right) - K_e \left(y + \Delta l \right)$$

$$+ K_z \left(\Delta z - y \right) + D_x \left(\frac{dy_c}{dt} - \frac{dy}{dt} \right) - D_e \frac{dy}{dt} - F \tag{2-88}$$

where $y, \dfrac{dy}{dt}, \dfrac{dy^2}{dt^2}$ — dynamic lift

range, velocity, and acceleration of valve

  m — system mass

  K_x, K_e, K_z — stiffness of system, spring and valve seat

  D_x, D_e, D_z — internal and external damping coefficient of system and damping coefficient of valve seat

  x_0 — valve clearance

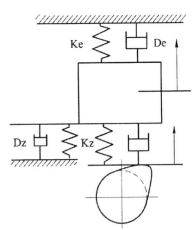

Fig.2-9 Dynamic model of valve-valve seat

Δl – initial deformation of valve spring

Δz – initial deformation of valve seat

F – the force acting on the valve from the chamber gas pressure

From equation (2-88), the valve dynamic performances including lift range, velocity and acceleration, can be obtained. Thus the stability and impact behaviour of the valve and valve seat can be evaluated.

(2) Bending stress of the valve head[11,49]

$$\sigma_w = \frac{0.12 p_z R^{(2m-0.4)}}{H^2\left(1-\lambda\gamma^{0.4}\right)}\left[\lambda\gamma^{0.4}\left(a^{2.4}-R^{2.4}\right)-\left(a^2-R^2\right)R^{0.4}\right] \le [\sigma_w] \quad (2\text{-}89)$$

where, p_z – maximum gas pressure

$$\lambda = \frac{(3m+1.6)(3m-v+0.6)}{(3m+2)(3m+v+1)}$$

m – coefficient with a range of 0.6~1.2, may be valued as 1.0 for

initial calculation

v – possion rate, 0.3 for steel

R, d, D, D_1, h_r – dimensions of valve head, referring to Fig.2-12(b)

$$\gamma = \frac{ad}{D_1+D}, a = \frac{D_1+D}{4}, H = h_r R^m$$

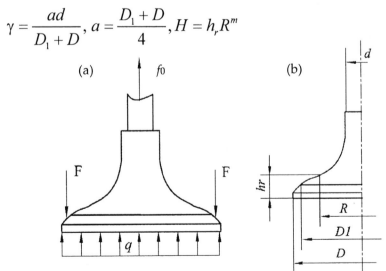

(a) (b)

Fig. 2-10 deformation and stress calculation of a valve head

(3) Deformation of the valve head[11,49]

The symmetry of the applied force and valve geometry, seen in Fig.2-10, simplified the deformation calculation of a valve head into a one-dimensional question. So only a part of the valve disk is analysed according to the bending deformation equation of a cantilever, i.e.,

$$\frac{d^2 y}{dt^2} = \frac{M(x)}{EI(x)} \tag{2-90}$$

$M(x)$ – bending moment at cross section x

$I(x)$ – inertia moment of neutral axis at the cross section x

E – materials elastic module of valve disk

Equation (2-90) can be solved to obtain the maximum axial deformation of valve disk based on the maximum gas pressure and the structure.

2.3.3 Criteria of acceptable design

Criteria of valve-tappet pair

(1) Critical hydrodynamic index Nr. At the cam nose region, the following equation should be kept.

$$0.15 \le N_r \le 0.25 \tag{2-91}$$

(2) Allowable Hertzian stress[σ_c]. The allowable contact stress is determined by the matching materials of cam and tappet. For the tappet with a punch head, there exists Table 2-8.

Table 2-8.   Allowable contact stress

Matching materials	Allowable stress
45# steel cam- 20# carburised quenching tapper	$[\sigma_c] = 60 \ kg \ / \ mm^2 = 600 \ Mpa$
45# steel cam – chill cast tapper	$[\sigma_c] = 65 \ kg \ / \ mm^2 = 650 \ Mpa$
Chill cast cam – chill cast tapper	$[\sigma_c] = 75 \ kg \ / \ mm^2 = 750 \ Mpa$

Criteria of valve-valve seat pair

(1) Dynamic properties of dropping movement of valve. Firstly, the dropping region should be at a transitional section (or a buffer section) of a cam profile curve. Secondly, the dropping velocity should be less than 0.5m/s for a normal valve seat made of cast iron.

(2) Allowable bending stress of valve head. For the generally used material croloy, $[\sigma_w] = 7kg/mm^2 = 70Mpa$

(3) Allowable deformation of valve head. The maximum axial deformation of the valve head should be less than 0.02mm.

2.3.4 Adjustable design parameters of tribological design

Adjustable design parameters for cam-tappet pair

(1) Basic circle radius of cam. Adjustment of the basic circle radius of a cam has a significant effect on the critical hydrodynamic index Nr and the contact stress σ_c, but has a minor influence on other performances of a valve system. Therefore, when the tribological design aim of a valve-tappet system is not satisfied, the basic circle radius r_0 becomes the first choice for a modifying design. A decrease of r_0 will result in an increase in contact stress σ_c and a decrease in the critical hydrodynamic index N_r, vice the versa. Therefore acceptable σ_c and N_r can be obtained by adjusting the value of r_0 during design.

(2) Cam profile curve. Modification of a cam profile curve also influences the magnitude of σ_c and N_r, however its design should rely on a comprehensive tribological analysis about the movement properties of the valve.

(3) Valve spring load. A large valve spring load results in an enhanced

load on the valve component, and thus causes increased wear and shortens the working life. While, a smaller valve spring load will lead to system disengage during working process, result in impact load and severe wear. Therefore, a proper selection is necessary for tribological design.

(4) Cam-tappet material matching. Different material used in the matching of cam and tappet determines different allowable contact stress values, which should be carefully selected according to the actual contact stress.

(5) Tappet structure. It has been demonstrated that, a tappet with a punch head corresponds with thicker oil film, which will result in lower contact stress and improve the tribological performances. Therefore, it is generally adopted in engines. However, as for a low-speed marine diesel engine, a tappet with a roller head is usually adopted so as that a rolling friction is realized instead of the sliding friction, which will decrease the frictional power loss and wear possibility.

(6) Material hardness and interface roughness. It has been reported that the relative material hardness between the cam and tappet should generally be less than 10 HRC. The hardness of a tappet should be larger than that of a cam. On the other hand, since the boundary lubrication state exists between the cam and tappet, the interface should be smooth enough to alleviate wear.

(7) Surface treatment. Surface treatment plays a significant role in the control of wear. Generally, a manganese phosphate treatment forms a compact and soft film on the surfaces of the matching pair, which will smooth the surfaces, help the friction pair avoid the severe wearing stage and transfer into a steady working period as quickly as possible.

(8) Lubricant and additive. Properly selected lubricant and additive will increase the film strength and the fatigue strength of contacting surfaces, and therefore decrease friction coefficient and wear.

Adjustable design parameters for Valve-valve seat pair

In case the tribological performances of valve-valve seat does not satisfy the design aim, a modifying design may focus on the following aspects.

(1) Dynamic dropping movement of valve. In order to implement a smooth dropping process, for a polynomial dynamic cam, polynomial index should be decreased, while for a circular cam, the basic circle radius of the cam should be decreased.

(2) Valve stiffness. When the stress or deformation of a valve head exceeds its allowable limits, that means the valve stiffness is not sufficient. The improving approach is suggested to be an increase of valve head thickness, back angle or the transition angle from the rod to the part.

(3) Material selection of valve and valve seat. As for the inlet valve, general carbon steel is a proper material for a light engine, and martensitic heat-proof steel for a heavy engine. As for the outlet valve, martensitic heat-proof steel is proper for a light engine, while austenitic heat-proof steel for a heavy engine. For a valve seat, alloy iron including Cr and Mo is proper for a heavy engine to increase the thermal stability.

(4) Valve seat hardness. The hardness difference between the valve and valve seat is suggested to be about 5HRc.

(5) Wear-resisting alloy is suggested to be welded on the valve cone to improve the wear-resistant performance.

(6) Oil fog is introduced into the inlet valve to keep the boundary lubrication state existing between the valve and valve seat in order to avoid the potential dry friction state which will result in a serious valve sinkage.

References

1. H. W. Hahn. Dynamically loaded journal bearings of finite length. Conference on lubrication and wear clut, *Mech. Eng.*, 1957

2. J. Holland. Beitrag zur ertassung der schmierverh ä ltniss in verdrennungskraft maschinen. VDI-Forsch-Heft 475,1959

3. J. F. Booker. Dynamically loaded journal bearings-mobility method of solution. *J.Basic Eng., Transactions of ASME: Series D*, 1965

4. Z.G. Qiu, C.S. Zhang. Analysis on a sliding roughness bearing with dynamic load. *Transaction of Internal combustion engine*, 1993,11(2),pp159-164(in Chinese)

5. X.L. Wang, S. Z. Wen and C. L. Gui. Lubricating analysis for dynamic load journal bearing considering surface roughness effects. *Chinese Journal of Mechanical Engineering*,2000, 36(1),pp27-31

6. R. S. Paranjpe. A transient thermohydrodynamic analysis including mass conserving cavitation for dynamically loaded journal bearings. *Transactions of ASME: Journal of Tribology*, 1995,117,pp369-378

7. X. L. Wang. Thermohydrodynamic analysis of engine main bearing considering surface topography effects. PhD thesis. Tsinghua University. Directed by Prof. S. Z. Wen and C.L. Gui, 1999

8. J. G. Jones. Grankshaft bearings: oil flow history. In Proc. 9th *Leeds-Lyon Symposium on Tribology, Tribology of reciprocating Engines*, 1982

9. K. P. Goenka. Dynamically loaded journal bearings: finite element method analysis. *Transactions of ASME: Journal of Tribology*, 1984, 106, pp429-439

10. N.Patir, H. S. Cheng. Application of average flow model to lubrication between rough sliding surface. *Transactions of the ASME: Journal of Tribology*, 1979, 101, pp220-230

11. C.L. Gui and J. Yang. Report of mechanical industry developing foundation item: study on design theory and method of tribology design of internal combustion engine. March, 1999, Hefei, China (In Chinese)

12. China State Bureau of Technical Supervision, GB1884~1885-80, China State Standard: *the conversion table of density experiment and measurement of petrol product*. Beijing: China Standard Press, 1980

13. B.X. Chen, Z.G. Qiu, H.S.Zhang. Hydrodynamic lubrication theory and application. Beijing, Mechanical engineering Presss,1991

14. G. A. Clayton, C. M. Taylor. Thermal considerations in engine bearing. *Proc. 17th Leeds-Lyon Symposium on Tribology*, 1990, 333-342

15. P.R. Yang. Numerical analysis of fluid lubrication. Beijing: National Defence Industry Press,1998

16. J.Ferron, J. Frene, R. A. Boncompain. A Study of the thermohydrodynamic performance of a plain journal bearing comparision between theory and experiments. *Transactions of ASME: Journal of lubrication Technology*, 1983, 105, pp422-428

17. R. Boncompain, J. Frene. Thermohydrodynamic analysis of a finite journal bearing's static and dynamic characteristics. *Proc. 6th Leeds-Lyon Symposium on Tribology*, 1979, pp33-41

18. M. M. Khonsari, J. J. Beaman. Thermohydrodynamic analysis of laminar incompressible journal bearings. *Transaction of ASLE*, 1987, 29, pp141-150

19. R. A. Castleman. A hydrodynamic theory of piston ring lubrication. *Physics*, 1936, 7, pp364-367

20. S. Furunhama. A dynamic theory of piston ring lubrication. *Bulletin of JSME*, First report-calculation, 1960(2),pp423. Second report - experiment, 1960(3), pp291-297. Third report-measurement of oil film thickness, 1961(4) ,pp744

21. T. Lloyd. The hydrodynamic lubrication of piston ring. *Proc. Inst. Mech. Engr.* 1968-1969, 183, pp28

22. D. Dowson, P. N. Economou, B. L. Ruddy, P. J. Strechan, A. J. S. Baker. Piston ring lubrication. Part II. Theoretical analysis of a single ring and a complete ring pack. *Energy Conservation Through Fluid Film Lubrication Technology*, 1979, pp23-52

23. S. M. Ruddy, P. N. Economou, D. Dowson. The theoretical analysis of piston ring lubrication and its use in practical ring-pack design. CIMAC, Helsinki, 1981

24. S. M. Rhode. A mixed frication model for dynamically loaded contacts with application to piston ring lubrication. Proceeding 7th

Leeds-Lyon Symposium on tribology, 1980, pp262-278

25. J. A. Greenwood, J. H. Tripp. The contact of two nominally flat surfaces. *Proc. Inst. Mech. Engrs.* 1971, 185, pp625-633

26. G. D. Knoll, H. J. Peeken. Hydrodynamic lubrication of piston skirts. *Transactions of ASME: Journal of Lubrication Technollogy,* 1982, 104, pp504-509

27. D. F. Li, S. M. Rhode, H. A. Ezzat. An automotive piston lubrication model. *Transaction of ASLE,* 1983, 26, pp151-160

28. D. Zhu, H. S. Cheng. A numerical analysis for piston skirts in mixed lubrication. Project Report, Northwestern University, 1991

29. Y. Z. Hu, H. S. Cheng, A. Jakayuki, K. Yoichi, A. Shunichi. Numerical simulation of piston ring in mixed lubrication. Project Report, Northwestern University, 1992

30. K. Liu. The investigation of the friction and Lubrication property of piston ring pack and establishment of wear model of ring-cylinder wall. PhD thesis, Xi'an Jiaotong University. Directed by Prof. Y.B. Xie and C.L. Gui, 1995, Xi'an, China

31. K. Liu, C. L. Gui and Y. B. Xie. A comprehensive investigation of the friction and lubrication properties of piston ring pack. *Tribology,* 1998, 18(1), pp32-38 (In Chinese)

32. K. Liu, Y. B. Xie and C. L. Gui. A comorehensive study of the friction and dynamic motion of the piston assembly. *Proc Instn Mech Engrs, Part I: Journal of Engineering Tribology,* 1998, 212, pp221-226

33. S. Furuhama, M. Takiguchi. Measurement of piston friction force in actual operating disel engine. SAE paper 790855, 1979

34. S. Furuhama, M. Takiguchi, K. Tomizawa. Effect of piston and piston ring designs on the piston friction force in disel engines. SAE paper 810977, 1981

35. S. Furuhama, S. Sasaki. New device for the measurement of piston friction force in small engines. SAE paper 831284, 1983

36. J. F. Archard. Wear theory and mechanisms. *Wear Control Handbook,* ASME, 1980, pp35-80

37. L. L. Ting. Lubricated piston rings and cylinder bore wear. Wear

Control Handbook, ASME, 1980, pp609-665

38. C. L. Gui. The Archard design calculation model and its application methods. *Lubrication Engineering*, 1990, No.1, pp12-21 (In Chinese).

39. K.Liu, Y. B. Xie and C.L. Gui. Two-dimensional lubrication study of the piston pack. Proc Instn Mech Engrs, Part I: Journal of Engineering Tribology, 1998, 212, pp215-220

40. D. C. Sun. A thermal elastical theory of piston-ring and cylinder bore contact. *Transaction of ASME: Journal Applied Mechanics*, 1991, 58, pp141-153

41. J. Holland. Die instationare elastohydrodynamik. Konstuktion, 1978, 30, H9, ss363-369

42. D. Dowson, C. M. Taylor, G. Zhu. Model lubrication of a cam and flat faced follower. *Proc. 13th Leeds-Lyon Symposium on Tribology*. 1986, pp599-608

43. X. Mei, Y. B. Xie. A numerical analysis for nonsteady EHL process of hign-speed rotation engine cam-tappet pair. *Transaction of CSICE*, 1994, 12, No.1, pp71-77

44. H.Q. Yu, X.Z.Zhan, Y.X. Huang. Elastohydrodynamics analysis of engine cam-tappet friction pair and the calculation of oil film. *Transaction of Internal combustion engine*, 1984, 2, No.2, pp125-140

45. G. Deschler, D. Wittmann. Nockenausheung für Flachstössel unter beachtung elastohydro-dynamischer schmierung. MTZ, 1978, 39, No.3, ss123-127

46. R. Müller. Der einfluss der schmier erhältnise am nockentrieb. MTZ, 1966, 27, No.2, ss58-61

47. M. Ryti. Zur rechnerischen optimierung von ventilsteuerunger. MTZ, 1973, 34, No.7, ss215-221

48. H. Yu, X. Zhan. The evaluation of *cam lubrication characteristics in I.C. engines.* Transaction of CSICE, 1983, 1, No.4, pp83-93

49. H. Guo, B. Zhuo, J. Peng, C. L. Gui. Study on valve-valve seat tribology designing. *Transactions of CSICE*, 2001, 19, No.3, pp258-262

Chapter 3: A Symbol Model-based Intelligent System and its Implementation

3.1 Introduction to the implementation of an intelligent system

An intelligent computer-aided system for engineering design emphasizes the cooperation between the human domain experts' design abilities and the calculating abilities of numerical analysis programs. In tribological design of an engine, the intelligent system aims to extend the various existing computer-aided analysis techniques (*e.g.*, lubrication analysis and wear prediction) as well as the abundant design knowledge (*e.g.*, design experience, standards, and manuals) into an intelligent design process. Three types of general knowledge models were introduced in Chapter 1 for the realization of an intelligent system: the symbol model, the artificial neural network model, and the gene model.

Two essential features are usually required for an engineering ICAD system: intelligence and integration[1,2]. To implement the intelligence feature, knowledge models will be developed to deal with different design decision-making problems. To implement the integration feature, an effective system architecture is needed to keep all the knowledge models, numerical analysis programs, and decision-making mechanism working seamlessly and efficiently.

In this chapter, an intelligent computer-aided system for engine tribological design (ICADETD) is established. First, the architecture of the system is proposed and generally illustrated. Then, the symbol models are described in detail as a basic knowledge representation, and object-oriented technology is used to implement the integrated system. Finally, an excellent example from normal design* of valve train system of engine is presented.

3.2 Architecture of an intelligent system for engine tribological design

The ICADETD system is made up of four main parts: an engineering knowledge bank and its administration mechanism, tribological analysis programs, inference mechanism, and an interface. The system architecture is shown in Fig. 3-1[3].

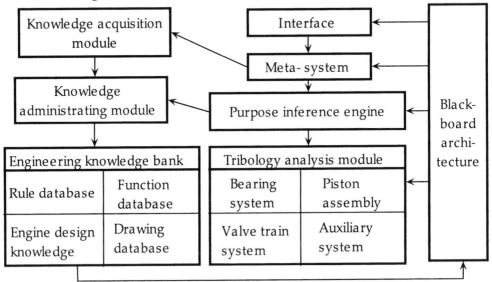

Fig. 3-1   Architecture of an ICAD system for tribological engine design

---

* Definition: If all the design parameters and variables referred to during the design process can be described as symbol models, and all the design strategies can be described step-by-step using the rules from experience and knowledge, then the design problem is assumed to be a "normal design problem".

Engineering knowledge bank. The engineering knowledge bank is made up of a rule database, an engine design knowledge database, a function database, and a drawing database. The rule database is a collection of design rules, prepared by domain experts, referring to the structural design, material selection, design criteria, and modifying design strategies. The engine design knowledge database includes various mechanical design standards and former design cases for parts and components. The function database includes numerical analysis tools, *e.g.*, finite element analysis software and engineering drawing software. The drawing database includes a variety of component and part drawings.

Knowledge administration module. The knowledge administration module is controlled by the purpose inference engine and is responsible for the operations of the database, including the searching, retrieving, inserting, editing, and modifying operations.

Knowledge acquisition module. The knowledge acquisition module serves as an interface between knowledge sources and the engineering knowledge bank, and acquires knowledge from domain experts or gathers various design data, in particular, the experimental and calculated data accumulated during earlier research into engine tribological design.

Meta-system and purpose inference engine. These two modules compose the decision-making mechanism of the intelligent system, control the design procedures, perform the design inference, deal with design conflicts, and eventually fulfill the design mission. Actually, they play an expert's role in the design process to manage data, invoke programs, make judgments, infer, and resolve conflicts. Therefore, they determine the intelligence level of the system. The working mechanism of these modules strongly depends on the knowledge representation methods used, *i.e.* the knowledge models. Different knowledge model adopted determines the different inference method and decision-making processes used. Therefore, they determine the ability of the intelligent system to deal with different design problems. The basic knowledge models of an intelligent system are symbol models, especially rules. Rule-based inference methods are usually forward reasoning or backward reasoning[4,5], and these form the essential methods of the purpose inference engine in the ICADETD system. The meta-system, meanwhile, mainly deals with higher-level inference process. This will be illustrated in the remainder of this chapter.

Tribology analysis module. The tribology analysis module is made up of powerful analysis techniques that employ finite element analysis, computational fluid dynamics, elastohydrodynamics combined with corresponding models of wear, fatigue and so on. It includes wear prediction of piston rings and cylinder liners, thermo-deformation of cylinder liners, analysis of the oil film thickness of bearing systems and valve trains, and frictional power loss calculations. Multi-discipline constraints are involved in these tribology analysis programs, including thermodynamic constraints, dynamic constraints, strength constraints, and environmental protection constraints. In the tribology analysis module, each design scheme can output its own simulated tribology performances as the basis of its own acceptability evaluation.

Blackboard architecture. The blackboard architecture is designed to be a global data structure, and serves as a dynamic information-sharing area, to record the input design mission from the user, and to present intermediate information and facts during the design process and eventually present the design results. Therefore, the blackboard architecture consists of a number of knowledge sources, which communicate with each other under the control of inference mechanism. The inference mechanism monitors the context of the blackboard, and judges whether or not the current data is suitable to trigger a knowledge model (*e.g.*, a rule). After a knowledge model is triggered, its execution results are listed on the blackboard to be evaluated once again by the inference mechanism. Details regarding blackboard architectures can be found in the references [6] and [7].

Interface. The interface controls the exchange of information between the user and the intelligent system, by which the user can input the design mission, monitor and participate in the design process, and submit advice.

3.3 Symbol models in engine tribological design

Knowledge representation is the first issue to be addressed in building an intelligent system. The nature of engineering design, and the diversity and complexity of engineering knowledge, requires knowledge representation to be flexible and robust. Owing to the domain characteristics of engine tribological design, the relevant knowledge involved is abundant, but trivial. Some domain knowledge are explicit, and can be represented easily, and some are implied in experimental data, in calculated data, and in earlier designs. Symbol models, including rule, framework structure and etc, play a significant role in representing the explicit knowledge.

Rule

Rule is the most commonly used symbol model to describe heuristic expert experience, judgment, and definition. It takes the following form.

IF(precondition is "TRUE")
THEN(conclusion)   **CF**(credit value)

The above conclusion may be a judgment or a calculation procedure. The credit factor (CF) mechanism has been adopted in our research to indicate reliability of the rule. The CF value (*i.e.*, the credit value) is suggested by domain experts. To cope with any rule conflict, several alternative rules can be automatically ranked by their credit values to determine their relative priority. An example of the CF mechanism is as follows

Rule 1:
     IF (frictional power loss of piston assembly > the permitted value)
     THEN (deduce width of piston ring using a step size of 0.1 mm with
          minimum =2.8 mm)   **CF** = 0.7.

Rule 2:
     IF (frictional power loss of piston system > the permitted value)
     THEN (decrease surface roughness of the cylinder line using a step
          of one grade to a minimum = 0.012 μm)   **CF** = 0.8.

If the two rules are activated simultaneously, then rule conflict occurs. Which rule should be enacted is determined by the inference engine, which uses the credit values and the current design parameters' suggested value fields. In our research, each design parameter owns its own field range suggested by the domain experts. In the above example,

Rule 2 is to be preferred to Rule1. However, if the design parameter involved in Rule 2 exceeds its suggested value field, then Rule1 will be adopted instead.

The CF mechanism is the most simple and direct conflict-resolving strategy. The CF mechanism itself is knowledge-intensive, and varies with the design mission. Therefore, most rule conflicts cannot be solved using only the CF mechanism. When a higher lever design conflict appears, the meta-system is activated to deal with it. The meta-system will be discussed in remainder of this chapter.

In the normal design problem defined earlier, all the design parameters and variables referred to during the design process can be described as symbol models, and all the design strategies can be described step-by-step using the rules gained from experience and knowledge. Therefore, rules can form an inference chain to guide the design process. For example, the inference chain for a design of a cam for a valve train system, in which the parameters P, Q, R, and S are the profile parameters of a four-power polynomial cam, is formed as follows

Rule 1:

    IF (kinetic or dynamic performance of the cam is not satisfied)
    THEN (decrease the value of P using a step length = 1 to a minimum
        value = 6)

Rule 2:

    IF (kinetic or dynamic performance of the cam is not satisfied) and
    (parameter P exceeds its value field)
    THEN (decrease the value of Q value using a step length = 2 to a
        minimum value = 10)

Rule 3:

> **IF** (kinetic or dynamic performance of the cam is not satisfied) and
> (parameter Q exceeds its value field)
> **THEN** (decrease the value of R using a step length = 2 to a minimum
> value = 20).

Rule 4:

> **IF** (kinetic or dynamic performance of the cam is not satisfied) and
> (parameter R exceeds its value field)
> **THEN** (decrease the value of S using a step length = 2 to a minimum
> value = 30).

It can be seen from the above that these rules form an inference chain, *i.e.*, Rule 1 – Rule 2 – Rule 3 – Rule 4. For a practical cam design problem, the inference chain described by parameters P, Q, R, and S constitutes a direct and efficient design inference path in the design space. Therefore, the purpose inference engine can perform the design process intelligently and automatically to deduce a successful solution in the design space.

The rules, CF mechanism, and inference chain are the basis of a purpose inference engine, by which the intelligent system can make design decisions and deal with simple rule conflicts in a direct and efficient manner.

Framework

A framework, as a knowledge representation method, is a structure that comprises nodes and their connections. A node is a named slot that owns its feature parameters and corresponding values[4], or owns a procedure written in a programming language[5]. Each slot can be developed to many new subnodes. Because it has a complex internal relationship, a framework is suitable for describing concepts and structures. A framework describing the concept of "bearing scuffing" is shown in Table 3-1.

Table 3-1. A framework describing the concept "bearing scuffing" in an engine.

Concept	Bearing scuffing			
Nodes (slots)	Structure	Lubricant	Materials	Working conditions
Feature parameters	Width, Diameter	Supplying amount, Additive, Viscosity	Properties, Surface topography	Pressure, Load, Speed

From the above model, it can be concluded that several parameters can be modified to avoid the scuffing of bearings, including improving the supply of lubricant, selecting the proper lubricant additive, improving the material surface topography, and changing the working load.

In addition to the rule and framework methods, object-oriented technology is considered a promising representation approach for engineering design problems. An object-oriented technique is defined as a software development strategy that organizes software as a collection of objects that contain both data and behaviours. Its application in the intelligent system for engine tribological design will be described in the following sections. It is a good strategy to combine all these symbol models to solve engine tribological design problems.

3.4 Working principles of the ICADETD system

As shown in Fig. 3.1, the blackboard block is the data centre of the ICADETD system, and the purpose inference engine and the meta-system make up the control centre. The working principles of the ICADETD system are illustrated in Fig. 3-2.

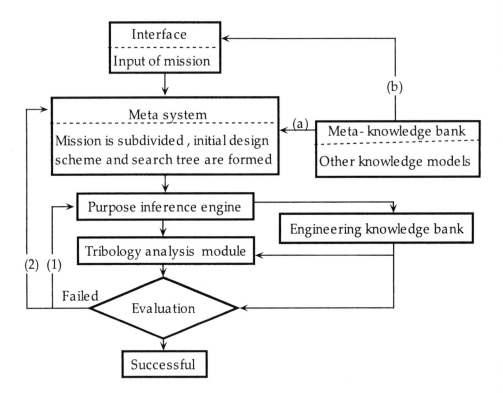

(1) Design is modifiable; (2) Rule conflicts
(a) Resovable conflicts; (b) Dissolvable conflicts, design fails.

Fig. 3-2. The working principles of the ICADETD system. The purpose inference engine and the meta-system make up the control centre of the system.

When a design mission is input into the intelligent system through the interface, it is firstly subdivided by the meta-system and sent to the purpose inference engine. Then, the relevant design knowledge, standards, and drawings are retrieved from the engineering knowledge bank by the knowledge administration module. After that, an initial design scheme is formed in the blackboard, and the tribology analysis module is invoked to simulate the corresponding tribology performances. Based on the analysis results, the initial design scheme is evaluated according to the design criteria to determine whether it is an acceptable design. If some of the predicted tribology performances do not meet the design objectives or constraints, then the current design scheme fails, and furthermore, its deficiencies will be listed in the blackboard. Based on these deficiencies, the intelligent system searches for matching design rules from the rule bank and automatically modifies the current design scheme. The revised design scheme is then sent to the blackboard, and is simulated by the tribological analysis programs once again. Under the control of the purpose inference engine, the evaluating and modifying process continues until a successful design scheme is eventually obtained that satisfies the design objectives and the tribological performance demands.

However, owing to its serial inference mechanism, the purpose inference engine is inclined to drop into a "dead-circle" when rule conflicts occur, or if several design parameters interact with each other. In this situation, the meta-system is activated to enact complex design decision by searching its own meta-knowledge bank or trigger other knowledge models. The complex decision-making process depends strongly on the knowledge models of the system. Among these potential knowledge models, the ANN model can extract symbol knowledge from engineering data to make up shortages in rules (see Chapter 4 this book), the gene model can provide the foundations for comprehensive parallel decision (see Chapter

5 of this book). A hierarchical gene model can obtain compatible solutions for a whole set of complicated systems and several subsystems (see Chapter 6 of this book).

Knowledge models in the intelligent system and their functions in design decisions are illustrated in Table 3-2. They provide a variety of knowledge representation methods and problem-solving strategies. Therefore, it is very important to make all the inference engines (rule-based serial inference, ANN-based parallel inference, and GA-based comprehensive parallel inference) work together to solve complicated engineering problems in which the meta-system plays an extremely important role.

Table 3-2. Knowledge models and their functions in design decision-making.

Function \ Model	Symbol model(rule)	Artificial neural network model	Gene model	Hierarchical gene model
Inference mechanism	Purpose inference engine	Purpose inference engine and Meta-system	Meta-system	Meta-system
Inference procedure	Serial inference	Supply rules for symbol models	Comprehensive parallel inference	Comprehensive decision-making for complicated system
Design problem	Normal design problem	Normal design problem	Non-normal design problem for complicated subsystem	Cooperative design for the entire set of complicated systems

It must be mentioned that to complement a design mission, the intelligent system is required to draw conclusions from engineering models, explain their results in terms of the design objectives and constraints, and to interact with the engineers in terms of the qualitative description of the system behaviour, rather than merely by numbers.

During the design process, the overwhelming volume of information produced by the numerical analysis programs raises its own problems in terms of evaluation and interpretation. A knowledge-based approach is very helpful in this situation. To aid interpreting the numerical results, the intelligent system should be able to[8]:

(1) Filter out (*i.e.*, recognize and represent) the important features of the large output data sets;

(2) Translate the features information at a level that will allow reasoning with and about them; and

(3) Make inferences based on the data representation concerning the state (*i.e.*, stresses, strains, failure criteria, and frictional power loss) of the object in question.

On the other hand, the numerical analysis output results are required to retrieve specific data, which are used to determine the design performance against the predetermined and established design criteria. The results may be in the form of stress values, displacements, or any predetermined data according to the design requirements. These data can be used in conjunction with the intelligent decision-making to determine the design acceptability.

As numerical analysis programs provide numerical data while the inference mechanism performs symbolic processing, it is necessary to convert the quantitative data of the analysis programs into qualitative statements for the intelligent system. For this purpose, we need to develop an "interpreter" that not only uses qualitative statements to describe the data from the numerical programs, but also transforms the inference mechanism's results into a useable form for the programs to identify.

An "interpreter mechanism" is, therefore, established in the intelligent system for engine tribological design. Based on the numerical results from tribological analysis programs, the design criteria, objectives, and constraints are retrieved from the knowledge bank to be compared with the results. The differences are listed in the blackboard, and then are translated by the interpreter into symbolic representations, *e.g.* "kinetic or dynamic performances of the cam is not satisfied", "maximum Hertzian stress is high", or "bearing scuffing probability is high". Then, based on the matching rules, *e.g.* "decrease the value of S using a step length $= 2$ to a minimum value $= 30$", the interpreter changes the corresponding parameter value directly. Therefore, the interpreter mechanism bridges the gap between the numerical calculation and the symbolic inference.

3.5 Implementation of the ICADETD system

The intelligent system for engine tribological design is required to provide the following functions:
(1) Integrate different knowledge-representation models and inferences;
(2) Process various types of knowledge and data to make decisions;
(3) Utilize different problem-solving strategies;
(4) Perform the decomposition and classification of deigned tasks automatically; and
(5) Implement communication between several operating systems.

To implement the above functions and capabilities, a software system is needed to integrate the various knowledge models, tribological numerical analysis programs, different design inference and decision-making process together, and perform the design mission both intelligently and automatically. Object-oriented (O-O) technology is a suitable tool for this purpose.

Object-oriented technology is defined as a software development strategy that organizes software into collection objects that contain both data and behaviour[9]. O-O is superior in the crucial aspects of readability, robustness, reusability, maintainability, and ease of extension and construction. O-O techniques can appear from operating-system design to programming languages, from computer graphic to system analysis, and from computer networks to user interface design.

O-O techniques not only describe the structural character of complex objects, but also encapsulate the behaviour characteristics of those objects[9]. This means that, for an internal combustion engine, both its design parameters and its tribology performance analysis programs can be encapsulated into a unified data structure (*e.g.*, the class structure of O-O technology). Similarly, for the intelligent system, both its knowledge models and its inference mechanisms can be encapsulated into a unified class structure. Therefore, O-O technology is an ideal software tool to implement the ICADETD system.

An object-oriented class structure is initially designed to describe an internal combustion engine from the viewpoint of tribological design. In the intelligent system, an engine class is built up as a basic class, and is inherited by four subclasses: bearing class, piston assembly class, valve train class, and auxiliary system class, as shown in Table 3-3.

Table 3-3. Object-oriented class structure of an internal combustion engine.

Basic class of an internal combustion engine			
Attributes:			
e.g. rotational speed, normal power, cylinder number, and stroke-bore ratio, etc.			
Function:			
e.g. lubricant selection program, frictional power loss calculation program, etc.			
Class1: Bearing	Class2: Piston assembly	Class3: Valve train	Class4: Auxiliary system
Attributes: Width, diameter, clearance, etc.	Attributes: Ring number, surface topography parameters, etc.	Attributes: Diameter of cam, strength of tappet, geometry parameters, etc.	Attributes: Bump parameters, air cleanliness, etc
	Function: thermo-deformation,		
Function: calculation of the centre orbit and film thickness, etc.	field analysis, calculation of frictional power loss, prediction of wear, etc	Function: Kinetic and dynamic analysis of valve train, wear calculation of cam.	Function: Calculation of lubricating properties.

Then, based on the symbol model, the rules are represented as a rule class. The rules in the ICADETD system are divided into three categories that correspond to the rules involved in the initial design, the modifying design, and the design scheme evaluation. Each type of rules is designed using a special inference engine, and each inference engine deals only with a small set of rules. Therefore, the inference mechanism is efficient. The rule-based inference engine is a very popular approach, and therefore the algorithm for rule-based inferences will not be discussed in detail here. The following provides a formalized representation of a rule class (written in the Borland C++ language).

Class Rule

```
{
//// Rule structure for the initial design
      CString PreD_Name[i];   //the name of parameter involved in the
                                    initial design
      float   PreD_Value[i];  //the corresponding value suggested by
                                    domain experts


////   Rule structure for the modifying design
      CString   Pre_Name;   //the reason for re-design
      CString   ReD_Name;   //the name of the parameter involved in the
                                  re-design
      float     ReD_Max;   //maximum limits of the parameter
      float     ReD_Min;    //minimum limits of the parameter
      float     ReD_Step;  //modifying step of the value
      int       flag;        //flag to indicate addition or subtraction operation

////   Rule structure for design scheme evaluation
      CString   Judge_Name[i];   //the names of the evaluation parameters

      float    Judge_Value[i];    //their corresponding ideal values;
}
```

Finally, four function classes are designed to encapsulate the different functions of the intelligent system and to implement it. They are a knowledge class, a design class, an inference class, and an interface class, as shown in Fig. 3-3.

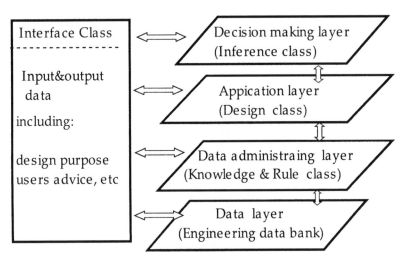

Fig. 3-3. Software structures of the ICADED system based on an object-oriented technology. The class is the basic structure and concept in O-O technology. The four main classes are designed with the corresponding tasks of interfacing, knowledge administration, numerical analysis, and decision-making.

The knowledge class includes the internal combustion engine class, the rule class described above, and others, which mainly represent the design knowledge and experience stored in the engineering data bank. The design class encapsulates the current design problem and the current design process. Inference class encapsulates the purpose inference algorithm and meta-system. The approach used to trigger other knowledge models and make complicated design decisions is also included in the inference class. The interface class encapsulates all the interfaces of the system that can perform data input and output functions. These four classes, working together with the engineering knowledge bank, form the object-oriented software structure of the intelligent system.

3.6. A case study of the intelligent design of a valve train system

Our laboratory has constructed an intelligent CAD system that can propose an initial design scheme, then predict its tribological performances, and evaluate the design scheme. When the design scheme does not pass the evaluation stage, this system can suggest modification strategies (taking the form as rules if a symbol model is triggered), and then suggest intelligent improvements to the design scheme. For a normal design problem, these relevant design rules can make up a complete inference chain, and can instruct the design process step-by-step. This procedure is maintained automatically until a successful and feasible design scheme is achieved.

As an example of this process, a design study from valve train system is presented to illustrate the ICADETD system. The design problem is described using design parameters, design objectives, and design constraints. The design parameters involved in a valve train system include cam type, cam basic circle radius, cam lift curve and so on. The multiple design constraints include the kinetic and dynamic property constraints, the cam maximum Hertzian stress constraint, and the valve head maximum axial deformation constraint, etc. (referring to Chapter 2). The tribological design objectives may include reducing the frictional power loss and preventing severe wear. These multiple design parameters, design constraints, and design objectives form a large and complicated design space. To search for an acceptable solution in such a large design space is not easy. However, in the intelligent design system, the search process can be guided by a combination of the experts' heuristic knowledge and various powerful analytical tools.

Our working example uses the 4105Q engine. An initial design scheme of the valve train system is formed from a former similar design case that was stored in the knowledge bank. Then, the intelligent system analyses the corresponding kinetic, dynamic, and tribological performances according to the design objectives and constraints, and evaluates them. In our design example, the initial design suggested has an improper cam type. Therefore, the purpose inference engine checks the relevant rules, and suggests a four-power polynomial cam. Then, the revised design scheme, including the profile parameters (P, Q, R, and S) of the polynomial cam is defined, and its corresponding tribological performances are analysed once again. If the revised scheme fails once more, then the purpose inference engine is activated and modifies the P, Q, R, and S parameters serially, according to the inference chain described above in Section 3.3. The inference process continues until a successful design scheme is obtained.

A screen snapshot of the design process in the ICADETD system is shown in Fig. 3-4, in which the suggested design parameters and the basic circle radius of the cam can be modified by the intelligent system according to rules, or directly by the user. However, for the user-modified route, the suggested values from the user should be within the value field specified by the intelligent system.

In summary, for a normal design problem, the design strategies form a decision-making tree that instructs the design process. For the example described, the selected and performed rules make up a "step-by-step ladder" in the design space defined by the multiple design parameters, objectives, and constraints. This "ladder" is very efficient and effective for the designer to find a solution. Therefore, for a normal design problem, the rule-based symbol model is the preferred selection mechanism for an intelligent system.

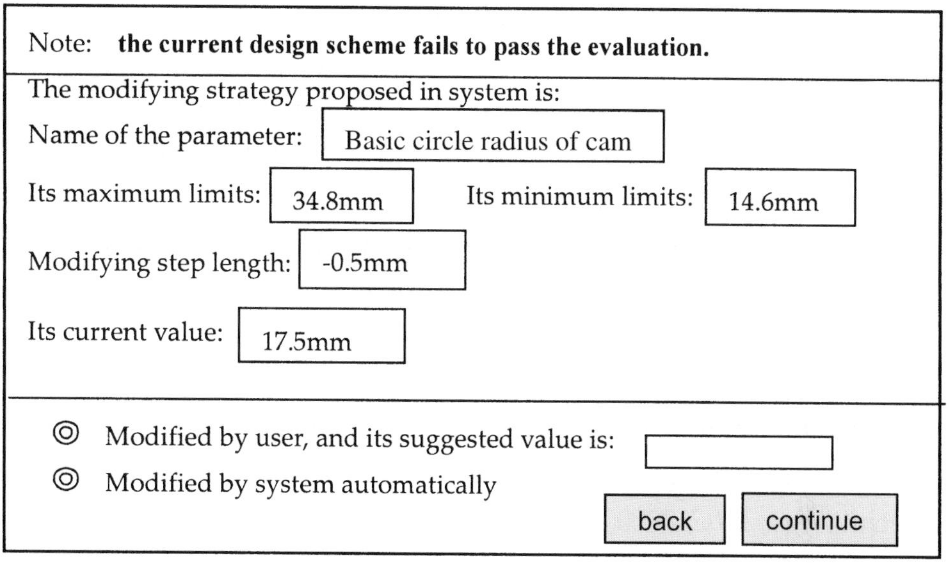

Note:	**the current design scheme fails to pass the evaluation.**

The modifying strategy proposed in system is:

Name of the parameter: | Basic circle radius of cam |

Its maximum limits: | 34.8mm | Its minimum limits: | 14.6mm |

Modifying step length: | -0.5mm |

Its current value: | 17.5mm |

◎   Modified by user, and its suggested value is: [   ]
◎   Modified by system automatically

[back]   [continue]

Fig.3-4. A screen snapshot of the interface in the ICADETD system. The parameter field range is listed besides its current value and modification step length.

3.7 Summary

Combining the inference ability of domain experts and the calculation abilities of numerical analysis programs, an intelligent system for engine tribology design has been developed that can solve tribological design problems intelligently and efficiently. As its basic knowledge model, symbol models are adopted to describe the domain design knowledge and the design inference process. To implement the intelligent system, object-oriented technology is used to integrate the knowledge models, numerical calculation programs, design inference processes, and decision-making procedures into a smooth and efficient operating software architecture. The intelligent system developed provides direct and efficient solutions to normal design problems encountered in engine tribological design.

References

1. V. Akman and T. Tomiyama. A fundamental and theoretical ramework for an intelligent CAD system. *Computer-Aided Design,* vol. 22, no. 6, July-Aug. 1990, pp 352–367

2. X. F. Zha, H. J. Du, and J. H. Qin. Knowledge-based approach and system for assembly-oriented design, Part II: the system implementation. *Engineering Applications of Artificial Intelligence.* Vol. 14, 2001, pp 239–254

3. X. J. Zhang. Study on the design methods of tribological design for internal combustion engine. PhD thesis, Hefei University of Technology, PRC, 2000, pp12

4. A. Hojjat. Knowledge engineering: fundamentals. *Springer Verlag,*1990

5. S. Tzafestas. Expert systems in engineering applications. Springer Verlag, Berlin, 1993

6. D. A. Sanders and A. D. Hudson. A specific blackboard expert system to simulate and automate the design of high recirculation airlift reactors. *Mathematics and Computers in Simulation,* vol. 53, 2000,pp 41–65

7. K. W. Chau and F. Albermani. Expert system application on preliminary design of water retaining structures. *Expert Systems with Applications,* vol. 22, 2002, 169–178

8. U. Roy. An intelligent interface between symbolic and numeric analysis tools required for the development of an integrated CAD system, *Computers in. Eng.,* vol. 30, No. 1, 1996, pp.13–26

9. Z. B. Chen and L. D. Xu. An object-oriented intelligent CAD system for ceramic kiln. *Knowledge-Based Systems,* vol. 14, No. 5-6, 2001, pp. 263–7

Chapter 4: The Artificial Neural Network Model and its Application in ICADEDT

4.1 Introduction to the artificial neural network model

Our former research emphasized that the core of an intelligent system for engine tribological design is its decision-making ability, which determines the overall intelligence of the system. In the intelligent system that we have developed, symbol models (rules, framework, and object-oriented representation) are used, and these can deal with normal design problems both efficiently and directly. These symbol models mainly represent the domain experts' heuristic knowledge and design experience. Among these, the rules are the most important models that can form a production inference chain for a normal design problem* and these can be carried out by inference mechanisms automatically. Therefore, the intelligent design ability of the developed system strongly depends on how many rules have been collected and stored, and how complete the formed inference chain would be.

However, there are many practical cases where the rules defining a problem are either not known, or are extremely difficult to discover. Such problems even exceed the field covered by the rule database, and several

---

* Definition: If all the design parameters and variables referred to during the design process can be described as symbol models, and all the design strategies can be described step-by-step using the rules from experience and knowledge, then the design problem is assumed to be a "normal design problem".

rules often conflict during a design process. Therefore, design strategies of such problems cannot be expressed completely using an inference chain, and the purpose inference engine in the intelligent system will drop into a "dead loop", owing to the shortage of higher-level design strategies. Hence, according to the working principles of the intelligent system mentioned in Chapter 3, the design problem or the conflict will be committed to the meta-system for it to seek for new rules, or trigger other new knowledge models.

Previous research into engine tribological design has garnered a great deal of valuable design information, which is implied in the abundant engineering data, such as calculated or experimental data. These information is difficult to express explicitly by domain experts or engineers. However, in order to enhance the intelligence of the ICADETD system, it is important to represent this implicit knowledge and to extract new rules from it.

As mentioned in Chapter 1, artificial neural networks (ANNs) are composed of elements or units that perform in a manner analogous to the neurones in the human brain. These units are interconnected through weighted arcs. ANNs can learn from training samples, and so reflect the implied relationships in their structures and weight values.

Much research[1-9] has demonstrated that the complex relationships implied in the abundant experimental and calculated data can be simulated by ANNs, and these relationships can be expressed explicitly by analysing the topography of the trained network[1]. ANNs, as a knowledge representation or acquisition tool in the field of engineering, have been applied to gear design[2], electrical machine design[3], radar signal analysis[4], and reliability prediction[5]. Clearly, an ANN will be an ideal solution to extract valuable knowledge from the abundant engineering data accumulated during the past research and practices, and will be significant for the tribological design in engines. However, research into ANN for use as a knowledge model in an ICAD system to deal with rule conflicts and make design decisions has not been reported.

4.2   BP net and its correlative variables

ANNs have been widely used in the engineering field in various structures and training algorithms. The feed-forward multi-layer perceptron net with back propagation algorithms, i.e. the BP net, is one of the most successful artificial neural nets, and was originally proposed by Rumelhart in 1986. A BP net is typically composed of three layers of simple units, as shown in Fig. 4-1. The BP net includes an input layer, i, a hidden layer, j, and an output layer, k. Each layer is fully connected to the previous layer through weighted arcs. Original data is presented to the input layer, processed through the intermediate hidden layer using non-linear Sigmoid functions, and finally, the network presents its results in the output layer.

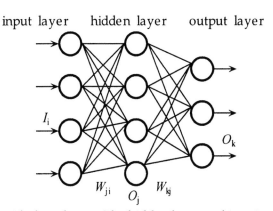

Fig.4-1. A BP net with three layers. The hidden layer and its trigger function play key roles in the simulation ability of a network.

In Fig.4-1, O_k is the output of the kth unit in the output layer, O_j is the output of the jth unit in the hidden layer, W_{kj} is the weighted value between the unit k and the unit j, W_{ji} is the weighted value between the unit j and the unit i in the input layer. n_k is the weighted input sum of unit k, and n_j is the weighted input sum of unit j. O_k and O_j are defined as

$$O_k = f_k(n_k) \tag{4-1}$$

$$O_j = f_j(n_j) \tag{4-2}$$

where f is the trigger function of neural unit. Usually, a trigger function in the hidden layer, f_j, is a Sigmoid function, as shown in Equation (4-3). Generally, no trigger function exists in the input layer, and the function in the output layer, f_k, is linear function.

$$f_j = \frac{1}{1 + \exp[-(n_j + b_j)]} \tag{4-3}$$

In the above equation, b_j is the variation and n_j is the input of the neural unit j.

To represent a relationship between the input layer and the output layer, the ANN learns from the training samples and adjusts its structure to fit the relationship implied in the training samples. The ANN's learning or training procedure is based on a simple idea: if the network suggests an incorrect output, then the weights on the connecting arcs are adjusted to reduce the output error.

To train a BP network, a group of training samples is required. Each sample comprises an input parameter part and an output parameter part. The samples are obtained from calculated data or experimental data, and their output is called the "actual output".

During the training process, each sample is presented to the network and propagated forward layer-by-layer until the output of the entire network is calculated. This output is called the "model output". Then, the model output is compared with the desired output, *i.e.* the actual output, and an error value is determined. The error signals are used to readjust the weights in a backward direction, from the output layer to the input layer. This process is repeated for all the training samples until the system error converges to a minimum, or until some specified number of iterations is reached. Finally, the model output will be close to the actual output, and the optimal weights are obtained.

The problem of training a BP network has been transformed into a non-linear optimization problem. The solution is to find the optimal weights that minimize the system error, which is defined as half the sum of the squared differences between the actual output and the model output at the output layer.

$$E = \frac{1}{2}\sum_{p=1}^{M} E_p = \frac{1}{2}\sum_{p=1}^{M}\sum_{k=1}^{N}(o_{pk} - x_{pk})^2 \qquad (4\text{-}4)$$

In Equation (4-4), O_{pk} and x_{pk} are the actual and model outputs for the kth node in the output layer for the pth training sample, respectively. M is the number of training samples, and N is the node number in the output layer. After training, the BP net is adjusted to form a proper structure and a group of arc weights, which make the error defined in Equation (4-4), stabilize in a preset limit.

In a post-trained BP net, the number of output nodes is usually defined by the application problem, and the number of input nodes is also application-dependent, because the nodes will usually correspond to objective features, *i.e.*, the independent variables. However, determining the number of hidden layers and hidden nodes is more complicated in a practical application.

Hormik *et al.*[1] have proven that standard feed-forward neural networks with a single hidden layer are capable of approximating any function from one finite dimensional space to another according to any desired degree of accuracy, provided that sufficient hidden units are available. Thus, a three-layer BP network with one hidden layer can be adopted in this work without any loss of generality.

Referring to Equations (4-1), (4-2), and (4-3), there is

$$f_j^{'}(n_j) = f_j(n_j)[1 - f_j(n_j)] \qquad (4\text{-}5)$$

For the linear function, f_k, there is

$$f_k^{'}(n_k) = a = const. \qquad (4\text{-}6)$$

In this case, the influence of unit i on unit k through unit j in the hidden layer is

$$\left(\frac{\partial O_k}{\partial I_i}\right)_j = \left(\frac{\partial O_k}{\partial n_k}\right)\left(\frac{\partial n_k}{\partial O_j}\right)\left(\frac{\partial O_j}{\partial n_j}\right)\left(\frac{\partial n_j}{\partial I_i}\right)$$

$$= f_k^{'}(n_k)W_{kj}f_j^{'}(n_j)W_{ji}$$

$$= a \times f_j(n_j)[1 - f_j(n_j)]W_{kj}W_{ji} \qquad (4\text{-}7)$$

The relationship between the input and the output is reflected in the weight values, which are invariable in a trained network. The value of n_j

can also be considered as being a constant for a group of inputs[10]. Therefore,

$$\frac{\partial O_k}{\partial I_i} \propto \sum_{j=1}^{s_1} W_{kj} W_{ji} \tag{4-8}$$

where s_1 is the number of neural units in the hidden layer. If we let

$$C_{ki} = \sum_{j=1}^{s_1} W_{kj} W_{ji} \tag{4-9}$$

then C_{ki} reflects the correlative relationship between unit i in the input layer and unit k in the output layer, *i.e.*, they are correlative variables. A positive value of C_{ki} indicates that the output increases as the input increases, and a negative value indicates that the output decreases as the input decreases. The value of C_{ki} indicates the magnitude of the influence. The correlative variables can be used to identify the influencing priority of the input parameters in a complex relationship.

4.3 ANN model in the ICADETD system

In a symbol model-based ICADETD system, rules can automatically constitute an inference chain, and instruct the design procedure directly and efficiently. This has been demonstrated by its application to the valve train system of an engine in Chapter 3 of this book. However, some complex non-normal design problems are characterized by a great many interacting parameters or constraints, and the relationship between the factors is often not known. In regard to this, an experts' corresponding experience may also be generally insufficient. Therefore, a new knowledge model needs be developed to represent such a complex relationship.

The application of an ANN model in the ICADETD system is shown in Fig. 4-2. To complete the ANN model's function, three steps are required:

(1) The training concept for the current problem, *i.e.*, what are the parameters involved in the input and output layers;

(2) Training the samples from the engineering data, *i.e.*, the calculated data or experimental data; and

(3) Analysis of the trained ANN model to extract useful information.

In general, both the training concepts and the analysis of the results rely on the knowledge of the experts. As a machine learning technique, ANNs are able to emulate human expertise or the relationships implied in the abundant engineering data, but they should never be allowed to replace human expertise.

In the intelligent system shown in Fig. 4-2, the symbol models initially supply the training concept, and the primary structure of the ANN model is established. Then, the numerical models (*e.g.*, tribological analysis programs), experimental data or former design cases supply the training samples to train the ANN model. After training, the ANN's structure and weights are transferred to the meta-system, which analyses the trained results, extracts new rules to strengthen the symbol models, and predicts the performance of the new, alternative numerical calculation and experiment.

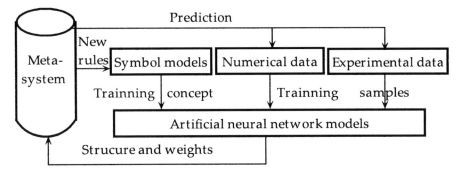

Fig.4-2.   Integration of an ANN in an ICADETD system

The following example discusses an application from a piston assembly as an illustration. In the tribological design of a piston assembly, there are about 10 parameters that can be modified simultaneously to decrease the frictional power loss, *e.g.*, widths of the piston rings, profile of the piston ring, the clearance between the piston ring and its slot wall, the surface roughness of cylinder liner and etc. To deal with the power loss problem, several rules have been proposed and stored in the ICADETD system, and these can suggest modifications to the piston assembly parameters.

IF (the frictional power loss > criterion)
THEN (deduce the surface roughness of the cylinder liner)

IF (the frictional power loss > criterion)
THEN (deduce the rotational speed of the engine)

IF (the frictional power loss > criterion)
THEN (deduce the width of the first ring)

Therefore, during the design process, if the precondition "the frictional power loss > criterion" is TRUE, then several rules will be triggered simultaneously. Despite the successful application of the CF mechanism in rule-based ICAD systems, the following question has not yet been answered, "How can a CF mechanism be automatically derived based on a given set of relevant rules?" Apparently, this question cannot be solved by the rules themselves, and the corresponding expert knowledge is also insufficient to determine the CF values, especially for multiple variable design problems. Therefore, the concurrent rules conflict with each other, the intelligent system cannot tell exactly which one is dominant, or most effective for the current design.

In this situation, any information about the effect on the frictional power loss of these parameters is extremely valuable for the design decision-making process. Although it is very difficult to be expressed explicitly, the information is believed having been implied in various former tribological analysis programs, including the Reynolds equations solution, thermo-deformation analysis, dynamic calculations, and wear prediction. The calculated data from these programs are believed to be valuable knowledge resources.

Therefore, an ANN model is built up to simulate the implied relationships. It has the structure as shown in Fig. 4-3. The input units represent several design parameters, which are, the average square root of the surface roughness of the cylinder liner, a, the width of the first compression piston ring, b_1, the width of the second piston ring, b_2, the width of the oil control piston ring, b_3, the normal rotating speed of the engine, n, and the lubricant dynamic viscosity, v. The output is the frictional power loss of the piston assembly system, w. The relationship between these parameters is described by the following equation,

$$f(a, b1, b2, b3, n, v) = w \qquad (4\text{-}10)$$

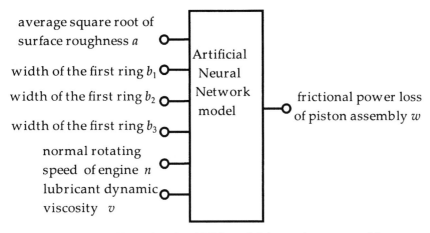

Fig.4-3.   An ANN model for a piston assembly

More than 20 groups of calculated data are obtained from a pre-developed tribological FEA program on the piston assembly, which calculates the frictional power loss, w, for a given input set of a, b_1, b_2, b_3, n, and v. Other design parameters are derived from the prototype design of 4105Q engine.

It is recommended that the input and output data must be normalized before presenting them to the network. Various normalization or scaling strategies have been proposed[8,9]. Scaling data can be linear or non-linear, depending on the distribution of the data. Most common functions are linear and logarithmic functions. A simple linear normalization function within the values of 0 to 1 is:

$$S = (V - V_{min}) / (V_{max} - V_{min}) \qquad (4\text{-}11)$$

where S is the normalized value of variable V, V_{min} and V_{max} are variable minimum and maximum values respectively. Therefore, each group of calculated data is scaled between 0 and 1 in a non-dimensional way to form a training sample similar to that shown in Table 4-1.

Table 4-1. Part of the training sample set formed for the piston assembly.

Parameters / Values	a	b_1	b_2	b_3	n	v	w
Sample1	0.8	0.98	0.86	0.95	0.98	1.0	0.96
Sample2	0.92	0.83	1.0	0.99	0.84	1.0	0.92
Sample3	1.0	0.95	0.87	0.95	1.0	0.96	0.97
Sample4	...	...	...	...	...	...	...

The architecture of the ANN has now been designed and tested, and this leads to six neural units in the input layer (representing the six parameters), 30 neural units in the hidden layer, and one neural unit in the output layer. The error limit of the ANN is set to 0.55×10^{-4}, and after training for 8495 epochs, the ANN model converges. The error-time analysis is shown in Fig. 4-4.

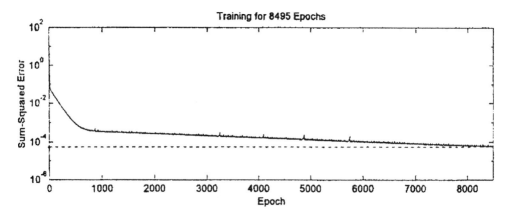

Fig. 4-4   Error-time behaviour of the ANN model during the training process.

During the application of the ANN model, the complex relationships between the six design parameters and the frictional power loss are realized by a set of weight values. It is expected that new heuristic information can be extracted from these to deal with any of the rule conflicts encountered above. It must be mentioned that the trained network for the current design is of the 4105Q engine. There will be different structures and weights for other design problems.

4.4   Analysis of the trained ANN model for piston assembly and its results

Up to this point, the extraction of symbol knowledge from the trained neural networks has turned out to be one of the most interesting open problems. Previously, some researchers have concentrated on the structure[2,3], and others have been interested in the hidden layer information[4]. In our research, to deal with the rule conflicts that emerge in the tribological design of a piston assembly, the priorities of the design parameters for the current design need be determined. Therefore, the correlative variables of the trained network should be considered and analysed.

According to Equation (4-9), the correlative variables of the trained network in Fig. 4-5 can be calculated as follows

$$C_{ki} = (2.1811, 1.7214, 0.5335, 0.8985, 3.2075, 3.0128)$$

This expression sequentially indicates the correlative values of the average square root of the surface roughness, a, the width of the first compression ring, b_1, the width of the second piston ring b_2, width of the oil control ring (the third ring) b_3, normal rotating speed of the engine n, and the lubricant dynamic viscosity, v. From the definition of C_{ki}, it can be seen that the engine normal rotating speed is the most important influencing factor on the frictional power loss, with the maximum correlative value of 3.2075. The second most important influencing factor on the frictional power loss is the lubricant viscosity, with a value of 3.0128. However, these are all the collective basic parameters for the engine, and should not be modified blindly during the piston assembly design process in order to avoid influencing the other subsystems of the engine, *e.g.*, the bearing system and the valve train system.

For local parameters, the average square root of surface roughness of cylinder liner has a higher correlative value of $a = 2.1811$, which means that modifying the variable a will play a more important role in reducing the frictional power loss. Therefore, a is the preferred parameter for modification design. Among the three piston rings, the width of the first ring is also the preferred option during the modification procedure, followed by the width of the third piston ring and the second piston ring.

Therefore, according to the CF mechanism illustrated in Chapter 3, and referring to the value fields of the design parameters stored in the ICADETD system, several new rules can be generated and ranked by different credit values, as shown in the following.

For the current piston assembly design
Rule 1:

> **IF** (the frictional power loss > criterion)
> **THEN** (deduce the surface roughness of the cylinder liner within
>   its value field)   **CF** = 0.9

Rule 2:

    IF (the frictional power loss > criterion)

    THEN (deduce the width of the first ring within its value field)

    CF = 0.8

Rule 3:

    IF (the frictional power loss > criterion)

    THEN (deduce the width of the third oil control piston ring within

        its value field)   **CF** = 0.7

For the entire engine set

Rule 1:

    IF (the frictional power loss > criterion)

      THEN (deduce the rotating speed of the engine within its value

        field)   **CF** = 0.9

Rule 2:

    IF (the frictional power loss > criterion)

    THEN (select low viscosity lubricant within its value field)

    CF = 0.7

Using these new rules, the priorities of the parameters that available to be modified during the modifying design stage may be determined. In these rules, each parameter has its limit of maximum or minimum value, which is suggested by the domain experts. If a parameter in the current scheme exceeds its value limits, then its corresponding rule will be ignored, and the rule with a lower credit value will be triggered sequentially. Therefore, the adopted sequence of rules, *i.e.*, the design strategies, can be determined for the current design and the rule conflict can be resolved.

However, the new rules regarding the collective basic parameters are not easy to use. To modify the collective basic parameters involves a compromise between several subsystems of the engine. The design scheme, including the collective basic parameters, should be compatible with several subsystems of the engine. To this end, new knowledge models are needed to design the collective parameters to satisfy all the design criteria from different subsystems and frictional pairs. This will be illustrated in the following sections.

After all, the ANN model plays a significant role in representing the quantitative relationship between its inputs and outputs, and furthermore, it acquires valuable information implied in the abundant engineering data. To deal with rule conflicts encountered in the tribological design of an engine, valuable new rules derive from the correlative variables. These new rules strengthen the ICADETD system and enhance its decision-making ability.

In addition, when a set of new piston assembly parameters within the domain covered by the training examples is input into the trained ANN mode, it is convenient for the ANN to be able to predict and calculate the frictional power loss directly, instead of performing the time-consuming numerical analysis programs.

4.5 Summary

The purpose inference engine will fail during the intelligent design process involving engine tribology, when several rules conflict and the corresponding solving rules are not adequate. Therefore, the meta-system is activated to invoke new knowledge models. Among these, artificial neural networks play a significant role in acquiring implied knowledge from the abundant engineering data. The BP net is a successful ANN model, and its correlative variables indicate the influence of the input parameters on the output parameters. In this chapter, an ANN model trained by the abundant calculated data accumulated from tribological simulation programs has been established for a piston assembly system. Based on the analysis of its correlative variables, valuable design rules can be extracted to determine the adopting sequence of the conflicting rules, and so the corresponding rule conflict can be resolved. Therefore, when integrated with the pre-existing symbolic knowledge model, the ANN model enhances the decision-making ability of the intelligent system for tribological design of engines.

References

1. K. Homik, M. Stinchcombe, and H. White. Multilayer feedforward networks are universal approximators. *Neural networks.* 1989, vol. 2, pp. 359–366

2. D. Su, M. Wakelam, and K. Jambunathan. Integration of a knowledge-based system, artificial neural networks and multimedia for gear design. *Journal of Materials Processing Technology,* 2000, vol. 107, pp. 53–59

3. V. L. Zakarian, and M. J. Kaiser. An embedded hybrid neural network and expert system in a computer-aided design system. *Expert systems with Applications,* 1999, vol. 16, pp. 233–243

4. J. F. Remm, and F. Alexandre. Knowledge extraction using artificial neural network: Application to radar target identification. *Signal Processing,* 2002, vol. 82, pp. 177–120

5. Y. B. Moon, C.K. Divers, H.J. Kim. AEWS: an integrated knowledge

-based system with neural networks for reliability prediction. *Computers in Industry*, 1998, vol. 35, pp. 101–108

6. L. C. Jain, and N. M. Martin. Fusion of neural networks, fuzzy sets, and genetic algorithms: industrial applications. 1999, CRC press, New York

7. A. S. Garcez, K. Broda, D. M. Gabbay. Symbolic knowledge extraction from trained neural networks: A sound approach. *Artificial Intelligence*, vol. 125, 2001, pp. 155–207

8. I. A. Basheer, and M. Hajmeer. Artificial neural networks: fundamentals, computing, design, and application. *Journal of Microbiological Methods*, 2000, vol. 43, pp. 3–31

9. M. Y. Rafiq, G. Bugmann, and D. J. Easterbrook. Neural network design for engineering application. *Computers and Structures*, 2001, vol. 79, pp. 1541–1552

10. R. J. Zhan, Y. X. Zhang, and Y. B. Xie. Application of an artificial neural net to acquire knowledge in tribology design. *Lubricant and Seal.*, 1997, vol. 2, pp. 719–723,(in Chinese)

11. C. L.Gui, J. Yang, et al. Research on Theories and Methods of Tribological Design of Internal Combustion Engine. *Research Report of the Mechanical Industry Developing Foundation*, 1999, Hefei, China

Chapter 5: Gene Model-based Comprehensive Decisions for Tribological design

5.1 Non-normal design problems and comprehensive decision-making

A rule-based ICAD system for engine tribological design (the ICADETD system) has been established in the previous chapters of this book. The intelligence level or the design ability of the system depends on its inference mechanisms (the meta-system and purpose inference engine) and on the adopted knowledge models. A normal design problem* can be dealt with efficiently and correctly by a symbol model-based intelligent system, in which the rule model is the essential and basic knowledge model. When rules conflict or parameter interaction occurs, and cannot be solved by the purpose inference engine, an artificial neural network (ANN) model is adopted to extract valuable rules from the knowledge implied in the abundant experimental or calculated data. This approach has been used to solve rule conflicts and enhance the intelligence level of the system. However, no matter either the symbol model or the ANN model is adopted, the design inference mechanism of the intelligent system is a rule-based serial process.

In a non-normal design problem, the design space is very complicated,

---

* Definition: If all the design parameters, variables referred to during the design process can be described as symbol models, and all the design strategies can be described step-by-step using the rules from experience and knowledge, then the design problem is assumed to be a "normal design problem".

and an engineering design problem is often of a difficult nature. In general, engineering design is a process to achieve feasible solutions for given objectives, while simultaneously satisfying certain constraints. Quite often however, the conflicting interactions of the design parameters, design objectives, and the design constraints are so complex that they are not fully appreciated by the domain engineers. Therefore, there is a dearth of corresponding modifying strategies, when the design process cannot be described explicitly step-by-step by a symbol model, and using a serial inference mechanism is an extremely inefficient method. To implement such an engineering design problem, the corresponding design inference mechanism should be a comprehensive parallel design decision process.

Non-normal tribological design of an engine is also a multi-purpose, multi-constraint, multi-parameter problem, in which the design parameters and constraint interactions are extremely complex. In an internal combustion engine, the heat transfer, noise, vibration, and the structural and material constraints interact very strongly with the tribological considerations. For example, to avoid "scuffing of bearing", the clearance between the journal and the bushing needs to be enlarged. However, this will cause increased vibration and noise, and therefore violate the dynamic constraint. So, one needs to know how to balance the scuffing-resistant constraint and the dynamic constraint, or, furthermore, how to compromise between the multiple design parameters and the multiple design constraints. This is a comprehensive parallel decision-making process, and cannot be dealt with solely by rules and a serial inference mechanism.

Based on genetic algorithms (GAs), a gene knowledge model is proposed in this book. The basic principles of GAs have been introduced in

Chapter 1, including four parts: encoding mechanism, fitness function, gene operators, and the controlling parameters. GAs perform a multidirectional search in the complicated design space, and provide a stable convergence towards a near-optimum solution in many types of problems. GAs have the following advantages compared with conventional optimization and search techniques:

(1) GAs search in a population of solutions for feasible solution group, but not for a single optimization solution. This is significant for engineering designs, and especially for a comprehensive parallel decision-making problem, because GAs can provide the decision-maker with a full picture of all the possible compromise solutions. This makes the decision process easier.

(2) GAs impose no restriction on the objective function. The objective function may be discontinuous, and GAs do not need any further information about the problem.

(3) Conventional optimization methods are well developed for continuous models, but typical engineering design parameters may be integers, discrete, or mixed-continuous. GA-based search methods can overcome this difficulty.

Similarly, the comprehensive design decision based on the gene model can also be considered as being a near-optimum solution search procedure in a complicated solution space defined by multiple parameters, multiple objectives, and multiple constraints. The gene model can encode multiple design parameters into a gene string (*i.e.*, a potential design solution), integrate multiple design constraints into a fitness function, and then evolve and modify the gene population (the potential solution population) in parallel to obtain a group of near-optimum solutions. The near-optimum solution group provides a foundation for the intelligent system to make comprehensive decision, and therefore simplify the decision-making process to choosing an acceptable solution in the evolved population.

5.2 Evolutionary design and gene models

Genetic algorithms have been widely applied in engineering, and have developed into evolutionary programming (EP) and evolutionary strategies (ES). GAs, EP, and ES are collectively termed, "evolutionary computation" (EC)[1]. When EC is used in engineering design, it is termed as "evolutionary design" (ED).

According to the encoding mechanism of the GA, the design samples are represented as gene strings (gene samples), by which the design process is transformed to an evolutionary process, and the evolved results are considered as being potential design solutions. A comparison between a conventional design and an evolutionary design is listed in Table 5-1.

Table 5-1. A comparison between a conventional design and an evolutionary design.

Evolutionary design (Gene model)	Conventional design
A gene string	A design sample
Gene string population	Designed sample group
Fitness function	Design objective or criteria
Fitter gene strings	Better design samples
Crossover and mutation operators	Combinations of solutions for new design samples

Much research has been carried out on evolutionary designs[2-8]. Most of this has dealt with design optimization problems. Dr Anna Throdon from MIT has transformed a design problem into a parameter optimization design problem under design constraints[3]. The fitness function was constructed from a combination of objective and constraint functions by using a "penalty function" mechanism. In engineering, ED has also been used in pressure vessel design, Belleville spring design, and hydrostatic thrust bearing design[5].

Engineering design is based mainly on numerical simulation analysis. However, most of the simulation analysis functions are discontinuous, non-differentiable, and include noise. Moreover, for practical design objects, the multiple purposes and multiple constraints often conflict during the design procedure. All of these contributions make it more practical to find improved and feasible solutions instead of the calculated optimum solution.

There has been some research work focused on design problems with a complex design space and various design parameters, in which feasible, improved solutions were required instead of the best-calculated solution. A research group from Germany's Dortmund University[6] has used ED in the shape design of a supersonic nozzle. In their research, the supersonic nozzle was subdivided into 300 segments, and hence the design samples evolved in a design space made up of 10^{300} nozzles. Although the final nozzle shapes were not the optimum solutions, the performance of the nozzle was greatly improved, and the newly evolved shapes exceeded the expectation of the human experts. Among these solutions, several feasible solutions were selected after carrying out a few experiments, and one of improved solutions is shown in Fig 5-1.

It can be concluded from the above examples that the gene model can represent various design schemes (structure, shape, and parameters) in gene strings, and then the gene strings can evolve in the complex design space to find improved, feasible solutions. Such solutions are also known as "non-dominated solutions". This is a very practical point in the application to a comprehensive design decision in engine tribological design.

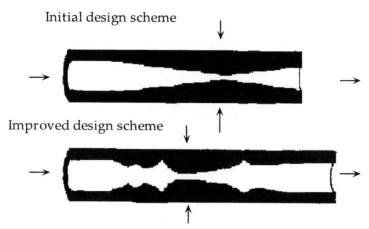

Fig. 5-1 One of the evolved design results of a supersonic nozzle.

5.3   A knowledge-embedded evolutionary design method

Evolutionary design has proven to be a suitable approach to search feasible solutions for multi-purpose, multi-constraint, and multi-parameter design problems. However, it is a blind search process, and is extremely time-consuming. To make the ED process more effective, advanced techniques have been adopted in sample representation, in evaluation mechanism, and in evolutionary operators. On the other hand, in the intelligent system for engine tribological design, abundant domain knowledge and design rules have been accumulated and stored. Rules can instruct a normal design procedure directly and efficiently, although they are insufficient to deal with a parallel comprehensive decision-making problem.

Therefore, we introduce a new knowledge-embedded evolutionary design method to combine the ED's comprehensive design abilities and the symbol models' heuristic instruction abilities, as shown in Fig. 5-2.

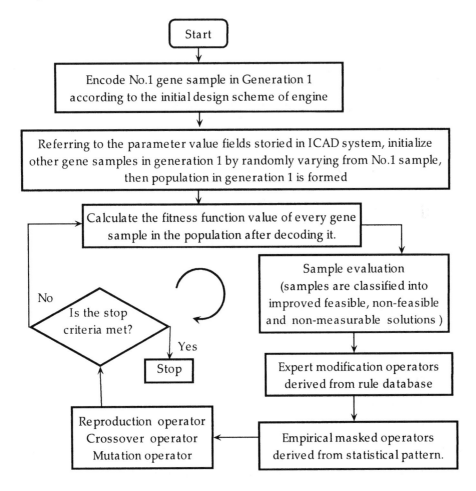

Fig. 5-2. A knowledge-embedded evolutionary design algorithm.

By denoting the population of a gene sample in a given generation as M(Generation), the flow of the knowledge-embedded evolutionary design algorithm can be summarized as follows in the knowledge-embedded evolutionary design algorithm (KEEDA)

KEEDA1:   Generation = 0
KEEDA2:   Initialize M(Generation)
              Evaluate M(Generation);
KEEDA3:   While (KEEDA has not converged or terminated)
        Generation = Generation+1;
        Reproduce feasible parts of M(Generation−1) to M(Generation);
        Crossover parts of the feasible solutions with parts of the
            non-feasible solutions in M(Generation−1), and keep the new
            offspring to M(Generation)
        Mutate parts of feasible and non-feasible solutions in
            M(Generation−1), and keep the new offspring to
            M(Generation)
        Modify parts of the non-feasible solutions in M(Generation−1)
            according to the experts' experience, and keep the new
            offspring to M(Generation)
        Mask parts of the non-feasible solutions of M(Generation−1)
            according to the empirical masked operators, and keep the
            new offspring to M(Generation)
        Evaluate M(Generation)
        End (While)
KEEDA4: Terminate the KEEDA.

The algorithms are characterized by a combination of the experts' heuristic knowledge with the evolutionary process. Taking the engine tribological design as an example, the KEEDA is described in detail as the following, in which design knowledge and experience are integrated in sample population initialization, sample evaluation, and newly developed evolutionary operators.

Encoding mechanism

In order to encode design variables into a binary string, we need to assume the length of the gene string, which depends on the required design precision, and the number of the design variables. In our research, each design variable is set to be a 3-bit gene segment. It corresponds with $2^3=8$ values in its domain field. Then, to represent a design solution, the entire length of a gene sample will has $3\times$(variable number) gene bits. Each 3-bit segment represents a design variable respectively. The corresponding variable values are derived from the interpolation in the value field, which are suggested by domain experts and previously stored in the knowledge bank of the intelligent system.

For example, the available lubricant viscosity grades are:

10W, 15W, 20W, 20, 30, 10W/30,5W/30,40

Therefore, the corresponding encoding mechanism for lubricant viscosity grade is,

3-bit binary segment	variable value
000	10W
001	15W
010	20W
011	20
100	10W/30
101	5W/30
110	30
111	40

In the proposed KEEDA, every design variable owns an interpolation table like the above. All the interpolation tables forms a bridge for the intelligent system to translate gene samples. A gene sample taken from the bearing system of an engine tribological design is shown in Table 5-2.

Table 5-2. A gene sample from the bearing system of an engine.

Gene model	110	101	100	011	111
Corresponding design variables	Viscosity grade of the lubricant	Width -diameter ratio	Width of the bearing (m)	Clearance (m)	Surface roughness (μm)
Corresponding variable values	40	0.35	0.028	0.0013	0.8

Gene population initialisation

As it was mentioned in Chapter 1, the initial population is typically generated at random. However, because the 4105Q internal combustion engine is regarded as being a design prototype, its design sample is regarded as No.1 design scheme in the initial population. So, other samples in Generation 1 derive from the initial design scheme of the 4105Q engine by varying different parameters at random in their predefined ranges *i.e.*, their value fields.

Sample evaluation

The sample evaluation mechanism is the core of evolutionary design, and it determines the evolutionary direction (*i.e.*, what type of design scheme will be selected out as a feasible solution). Owing to the multi-parameter and multi-constraint nature involved in engine tribological design, the treatment of the constraints is a main task in evolutionary design. We have unified various design constraints into a single constraint satisfying index, Y. Meanwhile, the frictional power loss is defined as the fitness function value, namely, the objective function value, M. In this way, the design problem becomes a bi-criterion problem, in which the first objective function is the sum of conflicting constraints, and the second objective function is the frictional power loss, which should be minimized. The design constraints for engine tribological design are illustrated in detail in Section 5.4. Based on the objective function value, M, and the constraint-satisfying index, Y, the sample population in each generation can be classified into improved feasible solutions, non-feasible solutions, and non-measurable solutions, which will be discussed in Section 5.5.

It must be mentioned that tribological analysis programs play a key role in predicting the tribological performance of each design sample. However, it is not proposed to develop the engine tribological analysis procedure any further here. The general principles and approaches have been analysed in Chapter 2 of this book.

During the design process, the experts' heuristic knowledge garnered from engineering practice is valuable to avoid arriving at meaningless solutions. For example, for decreasing the frictional power loss, the bearing diameter should be decreased. However, from the view of dynamics, a minimum limit must be satisfied, which is recommended by the domain experts. Therefore, only gene samples within their parameter limits have the chance to be evaluated as feasible solutions.

Termination criteria

As mentioned in Chapter 1, termination criteria is regarded as one of the control parameters of GAs. Generally, the maximum number of generations performed during evolutionary is set to be the simplest termination criteria. However, it seems to be better if the algorithm terminates the search when there is no more significant improvement in fitness function evaluations. For this purpose, there are two basic termination criteria. The first criterion measures the convergence of the population by checking the number of converged chromosomes. If the number of converged chromosomes is greater than some assumed percentage of the population, the search process will terminate. The second criterion measures the improvement in fitness function evaluation in a predefined number of generations. If the improvement of fitness function in the last generation is smaller than an assumed epsilon, the search process terminates.

In our research, the two criteria are combined together. Either of them is satisfied, the algorithm will terminate.

Improved Traditional operators

In addition to the simple one-point crossover operator and simple one-bit mutation operator presented in Chapter 1, the traditional operators are improved to enhance the effective of evolutionary process. Taking the two-point crossover operator as an example, it is implemented by choosing two or more points in the selected pair of gene strings and exchanging the sub-strings defined by the chosen points. The two-point crossover operator mixes information from two parent gene strings to produce offspring made up of parts from both parents.

Expert modification operator

An expert modification operator is constructed from rules that are stored in the ICAD system, which can modify "improper" gene bits in a non-feasible gene sample directly, according to the current conflicts. An example is illustrated in Fig. 5-3, in which if a gene sample is evaluated as being a non-feasible sample owing to its high scuffing coefficient, the rules stored in the intelligent system may suggest that "The distance from the chamber to the first compression ring should be increased, or the width of the first compression ring should be decreased". Then a temporary expert modification operator is formed, and this directly modifies the corresponding gene bits in the gene sample. Therefore, the expert modification operator can introduce the experts' heuristic knowledge to a GAs' blind search process, and enhance its research efficiency.

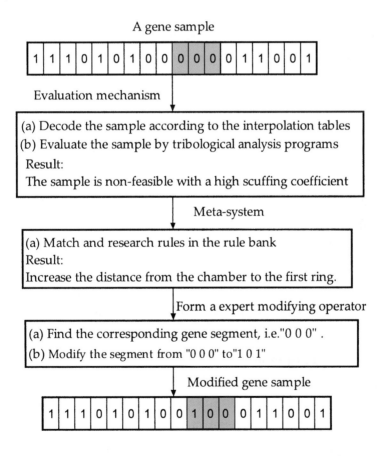

Fig. 5-3   An illustration of the working principle of expert modification operator.

Empirical marked operator

The empirical marked operator originates from the statistical results of the feasible solutions during the evolving process, and may be used to modify non-feasible gene samples. In our example, an empirical marked operator formed during the evolutionary design of a bearing system is 10**00**11*0*1*. This reflects a common statistical pattern of feasible solutions, in which the fixed binary bits (*i.e.*, the digits 0 and 1) represent the common bits owned by most of the feasible solutions, and the bits denoted by the symbol * reflect that there was no domain statistical results in those locations, and they may be arbitrarily set to 0 or 1. In reference to Table 5-2, the above operator denotes that the common features of the feasible solutions are lubricant of viscosity grade above 10W/30, a bearing clearance above 0.0013 m, and a moderate surface roughness. This reflects the commonly owned "good" gene segments in the feasible solutions.

When an empirical operator is acquired and activated, it masks on the non-feasible solutions. Only those non-feasible solutions owning the same segments will have more opportunities to be promoted to the next generation, and thus the corresponding "good" gene segments (or gene pattern) will be kept. Obviously, the process to obtain and apply the empirical marked operators is a process that the intelligent system learns itself from the feasible solutions, and adapts itself to the design problems.

All the above operators play an important role in the knowledge-embedded evolutionary design. During the evolutionary process, some of the feasible solutions are directly reproduced into the next generation; Some are promoted by the mutation or crossover operators together with the non-feasible solutions. Some of the non-feasible solutions are operated on by the expert modification operators or by the empirical marked operators. Non-measurable solutions are ignored. Using these operators, the proposed knowledge-embedded evolutionary method can combine the advantages of the ED and ICAD methods to deal with comprehensive design decision problems encountered in the tribological design of an internal combustion engine.

5.4 Constraints in the tribological design of an internal combustion engine

The tribological design of an engine is a multi-discipline, multi-constraint, multi-purpose, multi-parameter problem. In this book, the process is oriented at decreasing the frictional power loss and meanwhile satisfying the various design constraints. Because the design constraints derive from various disciplines and subsystems, and influence the evaluation standards and evolutionary direction, an important part of evolutionary design is how to deal with them.

Quite a large number of approaches have been developed recently to handle the constraints when ED are used[8], which can be roughly classified as follows:
(1) Rejecting strategy
(2) Repairing strategy
(3) Modifying genetic operator strategy
(4) Penalty function strategy

The last strategy has a universal feature. Therefore for most evolutionary design problems, constraints are regarded as being penalty terms of the fitness function[4]. If a constraint violation occurs, then the penalty terms will decrease the fitness function value through the coefficients of penalty terms. However, for an engine tribological design, since the design constraints derive from multiple disciplines, and take various forms, it is improper to combine these into a unified form of the penalty terms of the fitness function.

The main idea of the rejecting strategy is to discard all the non-feasible samples during the evolutionary process. This strategy works well when feasible samples constitute a reasonable part of the whole search space. In a practical design problem like engine tribological design, the initial population consists of mainly non-feasible samples in some cases, the strategy fails.

In the repairing strategy, a non-feasible sample is repaired to a feasible one using some repairing procedure. The weakness of such a procedure is that it can not be a universal procedure, and for each particular problem a specific repair method should be designed.

The modifying genetic operator strategy handles constraints by creating a problem-oriented representation of genetic operators to maintain the feasible of samples.

In our intelligent system for engine tribological design, the domain expert knowledge stored in the knowledge bank is transformed into the expert modification operator to repair and modify the non-feasible samples. The working principle is similar to that of the repairing strategy and the modifying genetic operator strategy. Meanwhile, the design constraints are regarded as a part of the evaluation mechanism, working together with the objective function, i.e. the fitness function.

Classification of design constraints

Design constraints that need to be addressed during the design procedure may be classified into three main groups,

(1) Parameter range constraints. For example, the stroke-bore ratio for a mini-type engine is restricted between 0.8 and 1.2. This type of constraint derives from the experts' design experience or from statistical results.
(2) Rigid constraints. These are design constraints that must be satisfied by the design samples. For example, the width of the piston ring and the width of its groove must maintain a consistent geographic relationship for them to be able to be assembled together.
(3) Soft constraints. These are constraints related to the engine's tribological performance. For example, the piston ring – cylinder liner frictional pair must satisfy the scuffing-resistant constraint[9]. Although scuffing is difficult to be predicted accurately only using numerical analysis, its relevant probability can be inferred and compared for different design samples.

The former two groups of constraints may be satisfied in the population initialization stage. The third group of constraints are very important in tribological design, and include the fuel consumption, dynamic properties, wear-resistant ability and so on. The concept of the constraint-satisfying index is proposed to denote the soft constraints.

Satisfying index of a single constraint

The satisfying index of a single constraint may be calculated according to the three distribution curves shown in Fig 5-4. These curves are

determined by the domain experts, according to the distribution laws of
the property parameters. For example, a design sample from a bearing
system presents a minimum film thickness, h_{min}, which must satisfy the
constraint shown in Equation (5-1).

$$h_{min} \geq [h_{min}] = 2(R_{zj} + R_{zb})$$

(5-1)

where R_{zj} and R_{zb} are the average surface roughness of the journal and the
bushing, respectively.

According to the distribution curve shown in Fig. 5-4(b), the
corresponding satisfying index, $y(h_{min})$, is defined in Equation (5-2).

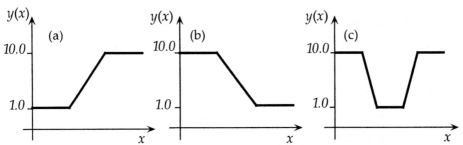

Fig.5-4 The distribution curves of a satisfying index of a single constraint.

$$y(h_{min}) = 10; when : h_{min} \leq [h_{min}]$$

$$y(h_{min}) = 1; when : h_{min} \geq 1.2 \times [h_{min}]$$

$$y(h_{min}) = 10 - \frac{10.0 - 1.0}{(1.2 - 1.0)[h_{min}]}(h_{min} - [h_{min}]);$$

(5-2)

In the above, the constraint-satisfying index takes its ideal value of 1.0,
and its value field has values of [1.0, 10.0]. The soft design constraints
considered in this book are summarized in Table 5-3.

Table 5-3. The soft constraints for an engine tribological design*

Bearing system	Scuffing-resistant constraint
	Highest temperature constraint
Piston assembly system	Scuffing-resistant constraint
	Leaking constraint
	Highest temperature constraint
	Piston knocking constraint
	Lubricant consumption constraint
Valve train system	Dynamic properties constraint
	Maximum Hertzian stress constraint
	Maximum valve head deformation constraint

*Reduction of the frictional power loss is set as the design objective of each subsystem

Comprehensive satisfying index for multiple constraints

The comprehensive satisfying index for multiple constraints is a combination of those of the single constraints, and takes two forms. One is the weighted sum of single-constraint satisfying indices, which is expressed by Equation (5-3).

$$Y = \sum_i a_i y_i \qquad (5\text{-}3)$$

where y_i is the satisfying index of a single constraint, and a_i is its weighted coefficient, as recommended by the domain experts.

Another combination method is shown in Equation (5-4):

$$Y' = 1/\left(y'_1 \times y'_2 \times y'_3 \times ... \times y'_n\right) \qquad (5\text{-}4)$$

where $y_1'...y_n'$ are the respective reciprocals of the single constraint satisfying indices with value fields of [0.1, 1.0]. It can be concluded from Equation (5-4) that only when every single constraint in Equation (5-4) is satisfied, the comprehensive satisfying index, Y', equals to its ideal value of $Y' = 1.0$. Therefore, Equation (5-4) is to be applied in subsystems where the single constraints are of the same importance.

5.5 Objective function and design sample evaluation

The engine tribological design problem considered in this book aims at obtaining improved feasible solutions based on the initial design scheme of the 4105Q engine. These improved feasible solutions should decrease the frictional power loss of the initial design scheme, while simultaneously satisfy all the design constraints described in Table 5-2. Therefore, an objective function can be defined as

$$M = W / W_0 \tag{5-5}$$

where W_0 is the frictional power loss of the initial design scheme, *i.e.* the frictional power loss of the 4105Q engine, and W is the frictional power loss of current design scheme evolved from the evolutionary process.

By now, each design scheme in the population has been evaluated using two functions: its objective function value, M, and its comprehensive constraint satisfying index, Y. Using these, solutions may be classified into three groups: improved feasible solutions, non-feasible solutions, and non-measurable solutions. According to the concepts of the objective function, M, and the constraint-satisfying index, Y, an improved feasible solution is defined as

$$x \in \{x_i \mid M_i \le 1.0; y_i \le 1.0\} \tag{5-6}$$

In the solution space described by M and Y, the improved feasible solutions are located in a rectangular zone contained within coordinates (0.0, 0.0) to (1.0, 1.0), called the feasible solution zone. This is represented by the hatched area in Fig. 5-5.

The samples with M or Y values greater than 10.0 are classified as non-measurable solutions, and are considered to beyond the scope of this book. The dots between the feasible and the non-measurable solution zones represent the non-feasible solutions.

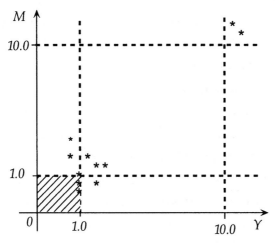

Fig. 5-5 Solutions in the solution space defined by M and Y.

Because of the complicated nature of engine tribological design, most of the design samples in the population are non-feasible samples at the beginning of the evolutionary process. Among these, those solutions having M values or Y values close to 1.0 are located near the feasible zone, and are considered to have valuable information and to be able to provide information to creative design. If they are ignored during the design process, the evolutionary algorithms incline to converge in a local optimum point. Therefore, how to utilize the non-feasible solutions is of great significance in the evolutionary design of an engine.

In this work, two new operators has been designed to utilize non-feasible solutions: the expert modification operator and the empirical marked operator. By modifying individual improper bits in a non-feasible solution (*i.e.,* the expert modification operator) or selecting quasi-feasible samples (*i.e.,* the empirical marked operator), valuable information included in the expert experience and in the non-feasible samples will be maintained during the evolutionary process and transferred to the next generation.

In this way, the objective function value, M, and the comprehensive constraint satisfying index, Y, are used to examine the potential solutions from different disciplinary points of view in parallel, and to evaluate every gene sample in the population. Then, the non-measurable solutions are neglected, and the feasible and the non-feasible solutions are evolved by applying various operators. Therefore, the average fitness value of the population is enhanced continuously until the entire population converges to a group of feasible solutions. To increase the design efficiency, the expert modification operator and empirical marked operator play key roles in utilizing the domain knowledge or valuable information contained within the feasible solutions.

5.6 Application to a bearing system and its results

In the evolutionary design of a bearing system, a gene model is built up as shown in Table 5-2. Then, an initial sample population with size of 20 is formed. A group of three-dimensional lubrication analysis programs is activated to calculate each sample's corresponding minimum film thickness, highest temperature, and frictional power loss. Using these, samples in the population are classified into feasible solutions, non-feasible solutions, and non-measurable solutions. Various operators are then activated and applied to the samples. The knowledge-embedded evolutionary algorithms are continually performed until the sample population converges finally at the 22nd generation, when the feasible solutions dominate the population.

(1) During the evolutionary procedure, the average objective function value curve of feasible solutions is as shown in Fig 5-6. This decreases as the generation number, t, increases, and eventually converges to a stable value of 0.85. This means that the minimum frictional power loss of the feasible solution is decreased to about 85% that of the initial design scheme, and cannot be further decreased randomly under the control of various design constraints.

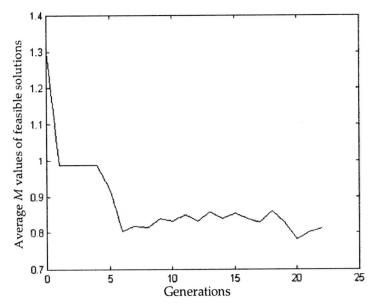

Fig.5-6. Average objective function value curve value of feasible solutions

(2) The average objective function value curve of the entire population is shown in Fig.5-7. This has higher values than that in Fig 5-6, and

converges to a value close to 1.0, which indicates that most of the samples in the population of the 22nd generation are feasible solutions.

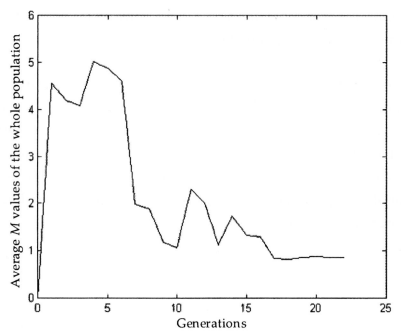

Fig.5-7. Average objective function of the entire population.

(3) In the solution space described by the objective function value, M, and the comprehensive constraint-satisfying index, Y, a design sample is represented as a point in the coordinate plane. During the evolutionary procedure, the solution space is as shown in Fig. 5-8 to Fig. 5-11. At the beginning of the evolutionary design, the sample points are scattered randomly in the solution space with few feasible solutions available. As the generation of evolutionary process increases, the sample dots gradually converge towards the feasible zone, and finally the feasible solutions dominate the population in the 22nd generation, as shown in Fig. 5-11.

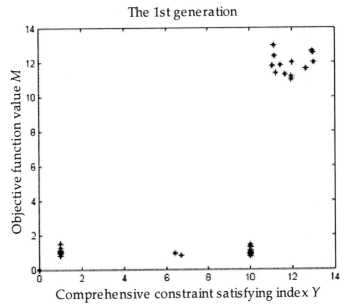

Fig. 5-8 Solution space distribution of the 1st generation during evolution.

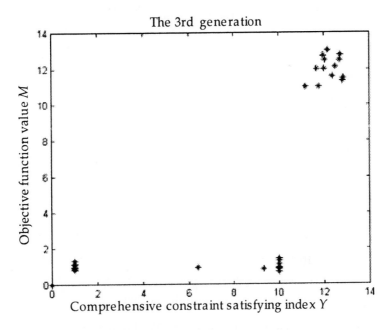

Fig. 5-9 Solution space distribution of the 3rd generation during evolution.

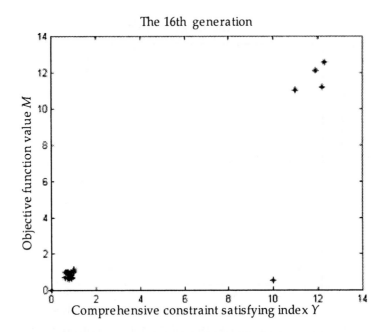

Fig. 5-10   Solution space distribution of the 16th generation during evolution.

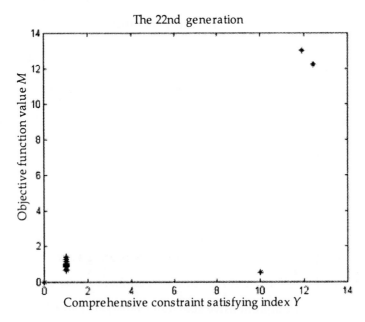

Fig. 5-11   Solution space distribution of the 22nd generation during evolution.

(4) After more than 20 generations' evolution, improved feasible solutions are obtained. Some of them are shown in Fig. 5-12, in which the y axis indicates an individual design scheme, *i.e.*, Scheme 1, 2, ... 7, and the x axis indicates the five design parameters and objective function value, *i.e.*, p_1 denotes the viscosity grade of the lubricant, p_2 the width–diameter ratio, p_3 the width of the bearing, p_4 the clearance between the journal and the bushing, p_5 the surface roughness, and p_6 the objective function value of the sample. Different line types represent different feasible solutions, in which several parameters are modified in parallel during the design, and finally become compatible with each other.

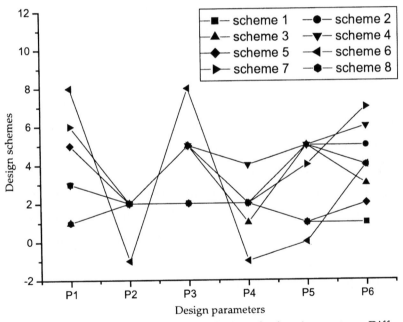

Fig. 5-12 Comparison of the feasible solutions for bearing system. Different line type represents different feasible solution, in which several parameters are compatible with each other.

The evolved feasible solutions, including several parallel design parameters, satisfy the design purpose as well as the multiple design constraints, and present the foundations for the meta-system to make comprehensive decisions.

5.7 Summary

The gene model-based evolutionary design approach modifies the design parameters in a parallel and compatible way during the design procedure. It then transforms the multi-parameter, multi-purpose, and multi-constraint design problem into an evolutionary procedure in the design space, and finally transforms the comprehensive decision into a selection process among the improved feasible solutions.

The proposed knowledge–embedded evolutionary design obtains feasible solutions under the instruction of domain experts' heuristic knowledge and the empirical knowledge acquired during the evolution. The expert modification operators derive from the design rules and the empirical marked operators derive from the statistical laws of feasible solutions. Both of them have successfully been applied to instruct the intelligent design and modify non-feasible solutions effectively and efficiently. The results of the knowledge–embedded evolutionary design make up the foundations for the meta-system to make comprehensive decisions in the ICADETD system.

References

1. M. A. Rosenman. An exploration into evolutionary models for non-routine design. *Artificial Intelligence in Engineering*. 1997, vol. 11, pp. 287–293

2. D. E. Goldberg. Genetic Algorithms in Search, Optimization, and Machine Learning. AWPC, 1989

3. A. C. Thornton. The use of constraint-based knowledge to improve the search for feasible designs. *Engineering Application Artificial intelligence.* 1996, Vol. 9, No. 4, 393–402

4. S. Esquivel, S. Ferrero, R. Gallard, et al. Enhanced evolutionary algorithms for single and multiobjective optimization in the job shop scheduling problem. *Knowledge-Based Systems*, 2002, vol. 15, pp. 13–25

5. D. Dasgupta, and Z. Michalewicz. Evolutionary algorithms in engineering applications. 1997, Springer-Verlag, Berlin

6. T. Baeck. An overview of evolution strategies. *International workshop on Evolutionary Computation.* April 2000, Wuhan, pp.166–201

7. D. A. Coley. An introduction to genetic algorithms for scientists and engineers. 1999, World Scientific Publishing, Singapore

8. A. Osyczka. Evolutionary algorithms for single and multicriteria design optimization. 2002, Physica-Verlag, Heidelberg, New York

9. C. L. Gui, and J. Yang. Study on the design theory and methods of tribological design of internal combustion engine. Report of the Mechanical Industry Developing Foundation, March 1999, Hefei, China (in Chinese)

Chapter 6: A Hierarchical Cooperative Evolutionary Design for a engines

6.1 Introduction to the design synthesis of subsystems and the entire engine

We have successfully applied the gene model to a bearing subsystem owing to its comprehensive decision-making capability, which mainly involves it in dealing with the local design schemes of a subsystem. However, if the focus of design process changes from a component to that of the entire set of the engine, then the synthetic design of such a complicated system is very difficult[1-3]. This is because it relates to more than one subsystems or components, and therefore, it needs a compromise among the subsystems. When the previously described gene model is applied in a design of the entire set of internal combustion engine including various subsystems, all the local design parameters involved in the bearing subsystem, the piston assembly subsystem, and the valve train subsystem are mixed together with the collective basic parameters of the engine. Then, they are encoded into a long "flat" gene string according to the established encoding mechanism mentioned in the former chapter.

However, we have seen that evolutionary design based on the traditional long "flat" gene string for complicated systems has some problems.

Parameter-dominating phenomena. Certain engine collective basic parameters dominate the fitness function value of the gene sample, which makes gene samples that include these basic collective parameters more likely to get higher reproduction possibilities and to be reproduced into the next generation. This prevents the non-dominant parameters (especially the local parameters of the subsystems) from evolving. For example, the engine rotational speed is a prime influencing factor on the engine frictional power loss, whereas the surface roughness of the piston cylinder liner plays a minor role. Thus, gene samples with lower rotational speeds will have a greater chance to be reproduced into the next generation, no matter what the value of the surface roughness is. This results in a group of feasible solutions with non-evolved surface roughness parameters. This kind of feasible solutions are not acceptable in practical designs. These phenomena are inevitably determined by the GAs' basic selection mechanism, *i.e.*, the reproduction possibility, Pr, introduced in Equation (1-6) of Chapter 1.

Subsystem-interference phenomena. Local parameters from different subsystems interfere with each other during the evolving process. A complicated machine is always made up of several subsystems that are usually non-interacting with each other and possess their own local design parameters. However, they are all connected with the collective basic parameters of the entire machine. For example, the surface roughness of the piston cylinder liner is unrelated to that of the bearing journal. During the evolving process, if a long "flat" gene string is made up of an excellent local design from the bearing subsystem and an infeasible local design from the piston assembly subsystem, then it will be evaluated as an infeasible solution, and therefore may be ignored by the selection mechanism. Thus, the excellent local design scheme for the bearing subsystem will be either modified (destroyed) or ignored. Although the destroyed design may be restored after many evolved generations according to Gas' principles, the design process will be extremely time-consuming, e.g., a three-dimensional lubrication analysis program of main bearings will take about 60 minutes on a 233 MHz PC .

J.S. Gero[3] also found that subsystem interference phenomena during his research on architecture layout designs, *i.e.* when represented as a "flat" gene string, the evolved rooms were always destroyed by non-evolved rooms within the same apartment. Therefore, he introduced a partition concept (*i.e.*, a small apartment) instead of considering the whole apartment.

In our evolutionary design of an engine, the parameter-dominating phenomena and subsystem-interference phenomena are extremely outstanding, because the local design parameters from a bearing subsystem, a piston assembly system, and a valve train are always unrelated. Meanwhile, they are easily dominated when mixed with the collective design parameters. Therefore, for an evolutionary design of the entire set of an engine, a long "flat" gene model is not a proper knowledge model, although it can work effectively within a subsystem.

To deal with complicated design problems, gene models designed with a complex structure and corresponding complex algorithms have been an interesting research issue in the field of evolutionary design[3-7]. However, few of these studies are aimed at such a complicated machine as an engine. To deal with the complicated machine like the entire engine, a granularity concept is first adopted in the chapter, by which the long "flat" gene model is transformed to a hierarchical gene model with a cooperating centre, then the corresponding new algorithms and cooperating mechanisms are established. Finally, the hierarchical gene model and its cooperating algorithms are used to the tribological design of the entire set of the engine.

6.2 Granularity of a complicated system

Generally, the evolving mechanisms of a complicated system can be summarized into three classifications:

(1) Parameter evolution with a fixed structure, *i.e.* the structure of the complicated system is pre-defined. The describing parameters are also pre-assumed, while the parameter values are viable and can be modified and improved during the evolutionary process. This is the normal case of Evolutionary Design (ED) process described in Chapter 5, i.e., the structure of the gene sample for the bearing system is fixed, and the design process is to find the compromising combinations among the parameter values.

(2) Iterative evolution along different structure layers of the complicated system. This evolution mechanism is similar to that of the human brain. The brain's basic structure or functional partitions are primarily determined when a person is born. Then, as the person grows up, the details of each functional partition are developed. New connections among partitions emerge and become strengthened.

(3) Structure evolution, *i.e.* the system is made up of a few of simple units in the beginning, and then, during the evolutionary process, new units are brought into the system. This enlarges the system, making it more complicated. For example, in a robot football game, if each football team includes only one member, the football game will be easy to control, since the competing strategies are easy to formulate. When the members of both teams increase to more than two or three players, the football is grabbed by several robots during the game. To search a proper competing strategy will be extremely complicated for each robot, since it requires cooperation or competition with other members.

Referring to the second evolution mechanism, a concept of granularity is proposed to describe the complicated system[8]. Granularity is defined as being the descriptive layers of a complicated system from the viewpoint of the design. Finer design models correspond to a higher granularity system that includes multiple layers and more details, and coarser design models correspond to a lower granularity system that includes fewer layers and details.

According to the definition of granularity, a hierarchical cooperating evolutionary design model is built up for a complicated system, as shown in Fig. 6-1. In this model, the granularity layer 0 is set as the centre, and includes a single unit, while other layers may include more than one subunits. Granularity reflects the distance of each layer (which includes more than one units or subsystems) from the granularity layer 0. Each unit can be expanded to several subunits (subsystems) again, which make up a new higher granularity layer.

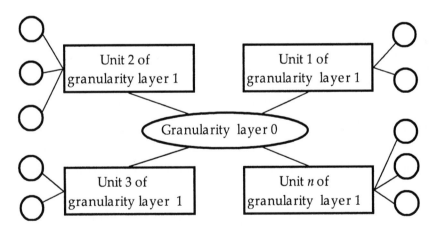

Fig.6-1. Hierarchical cooperative evolutionary model of the complicated system. The granularity layer 0 with only one unit is centre, while the granularity number represents the distance of the higher granularity layers to the centre.

In granularity layer 0, the design model for the complicated system is simple, and includes few parameters, owns few details and simple constraints. As the granularity increases, the design model expands and describes more details gradually, and becomes to be more exact. New design parameters and new constraints emerge, and the relationships between the design parameters and the constraints become increasingly complicated.

6.3 A hierarchical cooperative evolutionary design model for an engine

Taking the tribological design of an engine as an example, the design sample in the granularity layer 0 corresponds to that of the entire sample set, which is made up of the collective basic parameters of the engine, such as the rotational speed, the stroke–bore ratio. Its corresponding objective function (*i.e.*, simulating function of the frictional power loss) and constraints are relatively simple, which are listed in Appendix A of this book.

When expanded to granularity layer 1, the engine is divided into four subsystems (*i.e.*, the main bearing subsystem, the big end bearing subsystem, the piston assembly subsystem, and the valve train subsystem), in which the lubrication status of the 10 friction pairs listed in Chapter 1 must be taken into consideration. New parameters emerge in each subsystem, such as the widths of the piston rings, width–diameter ratio of the bearings, cam basic circle diameter, and the clearance between the valve and its guide. The corresponding objective functions are more complex, and there are various design constraints that must be satisfied, including the kinetic and dynamic constraints, the thermal constraints, and the maximum Hertzian stress constraints, etc., which have been described in detail in Chapter 2. Therefore, the design model of the engine as described in granularity layer 1 is more exact, but more complicated.

Meanwhile, as the granularity increases, more details emerge. For example, the surface topography parameter may be required to be expanded from a single parameter (*e.g.*. the average square root of the surface roughness) to several design parameters (*e.g.* the roughness height and the roughness orientation angle) to improve the accuracy of the friction models[9]. Accordingly, the more detailed descriptive parameters and the stricter constraints make the design problem increasingly complicated.

In the model described in Fig. 6-1, the normal evolutionary design method based on a "flat" gene string is adopted in each unit to obtain feasible solutions for each subsystem independently. To synthesize the design results from each unit, the granularity layer 0 is assumed to be the cooperative centre of the complicated system. Because excellent local solutions do not always make excellent solutions for the set of the entire system, a cooperative mechanism is required to harmonize the local solutions obtained from the different subsystems. Under the control of the cooperative mechanism, the local evolved solutions harmonizing with those from other subsystems will be adjusted to attain a higher fitness function value. Then, the local solutions return to their original units and evolve into the next generation. In the next generation, they are harmonized by the cooperative mechanism once again. This process repeats until feasible solutions for the set of the entire system are obtained that not only satisfy the design demands of their original units, but are also compatible with each other. The details of the cooperative mechanism will be introduced in the following sections of this chapter.

In summary, a hierarchical cooperative evolutionary design is an iterative evolving process along the granularity layers. Its aim is to obtain improved feasible solutions for the whole system. During the procedure, the design model evolves, the design parameters and constraints evolve, and the algorithms in each node involve evolutionary computation.

6.4 Parameter cooperative mechanism in the hierarchical system

For a hierarchical evolutionary model of an internal combustion engine, the parameters in granularity layer 0 are the collective basic parameters of the engine that influence the performances of all the subsystems. In contrast, the design parameters involved in higher granularity units are local parameters, which are normally unrelated to those of other units in the same layer. During the evolutionary process, the collective basic parameters make up the entire set of samples that are evolved in the coarse granularity layer with local parameters of subunits fixed, while the local parameters make up different local samples that are evolved in different subsystems in the finer granularity layers with the collective basic parameters fixed. From the system's point of view, both the solutions from the coarse granularity layer and those from the fine granularity layers must be harmonized with each other.

On the other hand, the results of the evolutionary design in each unit form a solution population instead of a single solution. Therefore, a population-based cooperative mechanism is required to harmonize these solution populations.

During an engine tribological design, the only commonly included parameter in the granularity layer 0 and the various subsystems is the lubricant. Therefore, the lubricant may be considered as being the cooperative basis for different solution populations. Thus, the population-based parameter cooperative mechanism proposed in our research can harmonize the solution populations obtained from both the granularity layer 0 and the different subsystems as follows:

(1) The design scheme from the prototype 4105Q engine is set as an initial design in each unit, and then the initial population is generated from it optionally and the ED routine is performed in each unit.

(2) After a preliminary evolutionary design has been carried out in each unit within the model, local samples from the different subsystems are initially classified into different groups according to their cooperating parameter, *i.e.* the lubricant. Those samples owns the same lubricant are grouped and labelled as a "local sample group".

(3) In granularity layer 0, if one of the entire set sample is compatible with a local sample group (with the same lubricant), then it is evaluated with each local parameter fixed at the parameter's average value over the its "local sample group". Then, it is set to have a higher fitness function value. If the entire set sample has no compatible local sample group, then it will be evaluated with each local parameter fixed at the parameter's average value over all the local samples, and its fitness function value will not be adjusted.

(4) During the evolutionary design of every subsystem, the entire set of samples is also grouped according to the lubricant. Meanwhile, each collective basic parameter is fixed as different average values over its different groups. If a local sample can find a compatible entire set of sample, it will be given a higher fitness function value. If not, the local sample is arranged a lower fitness function value.

(5) Returning to Step (2), the cooperative evolutionary process continues until feasible solutions dominate the sample populations of every granularity layer.

Therefore, the design process is an iterative procedure, moving along the various granularity layers, during which the entire set of samples and the local samples are continuously modified and evolved, based on their cooperating parameter, until feasible and compatible solutions for the whole engine are obtained.

6.5 Mathematical models of the hierarchical cooperative evolutionary design

(1) The design sample based on the granularity layer 0 model (*i.e.*, the entire set of samples) is assumed to be

$$\{x_1^0, x_2^0, x_n^0\}^T = \begin{Bmatrix} x_1 \\ x_2 \\ \vdots \\ x_n \end{Bmatrix}^0 = X^0$$

and the corresponding objective function is

$$f_0(x_1^0, x_2^0, \cdots x_n^0) = W^0 \tag{6-1}$$

The rigid constraints and soft constraints (in reference to Chapter 5 of this book) are

$$g_0(x_1^0, x_2^0, \cdots x_n^0) = 0 \quad \text{and} \quad h_0(x_1^0, x_2^0, \cdots x_n^0) < 0.$$

Therefore, the design problem becomes to be an optimization problem (to obtain minimum W^0) under these constraints. It is easy to obtain feasible solutions of X_{optim}^0 because of the fewer parameters involved, and the simpler objective function and constraints.

(2) When the model is developed to include granularity layer 1, several subsystems (nodes) are expanded. New design parameters and constraints are introduced in each subsystem

$$\left\{\begin{matrix} y_{11} \\ \vdots \\ y_{1n} \end{matrix}\right\}^1, \quad \left\{\begin{matrix} y_{21} \\ \vdots \\ y_{2n} \end{matrix}\right\}^1, \quad \left\{\begin{matrix} y_{31} \\ \vdots \\ y_{3n} \end{matrix}\right\}^1 \quad \dots ,$$

which are denoted by Y_1^1, Y_2^1, Y_3^1, and correspond with Subsystem 1, Subsystem 2, and Subsystem 3. Together with the entire set of samples, X^0, they make up the hierarchical gene model of the complicated system.

The objective functions of the model are

$$\begin{cases} f_{11}(X_0, Y_1^1) = W^{11} \\ f_{21}(X_0, Y_2^1) = W^{21} \quad \text{and } f(W^{11}, W^{21} \dots) = W^1 \\ \vdots \end{cases} \quad (6\text{-}2)$$

which are more precise than Equation (6-1).

The constraints are made up of two parts. Constraints derived from the granularity layer 0 model are considered to be

$$g_0(x_1^0, x_2^0, \cdots x_n^0) = 0 \quad \text{and} \quad h_0(x_1^0, x_2^0, \cdots x_n^0) < 0$$

and the newly emerged constraints include the entire set of samples.

$$g_{11}(Y_1^1, X^0) = 0, h_{11}(Y_1^1 X^0) < 0$$
$$g_{21}(Y_2^1, X^0) = 0, h_{21}(Y_2^1, X^0) < 0$$
$$\cdots$$

$$(6\text{-}3)$$

Then, the design problem is developed to a problem with multi-constraints, multi-parameters and complicated objective functions, for which it is not so easy to find optimum solutions.

(3) The cooperative evolutionary algorithms for the two-layer model is a three-step iterative procedure.

Step 1:

(a) X^0_{optim} is set to the initial design scheme of the granularity layer 0 unit.

(b) In the granularity layer 1, the whole set of the samples is fixed to X^0_{optim}, and the design problem is divided into several sub-problems:

$$\begin{cases} f_{11}(X^0{}_{optim}, Y_1^1) = W^{11}, \\ g_{11}(Y_1^1, X^0{}_{optim}) = 0, g_0(X^0{}_{optim}) = 0, \\ h_{11}(Y_1^1, X^0{}_{optim}) < 0, h_0(X^0{}_{optim}) < 0. \end{cases}$$

$$\begin{cases} f_{21}(X^0{}_{optim}, Y_2^1) = W^{21} \\ g_{21}(Y_1^1, X^0{}_{optim}) = 0, g_0(X^0{}_{optim}) = 0, \\ h_{21}(Y_1^1, X^0{}_{optim}) < 0, h_0(X^0{}_{optim}) < 0 \end{cases}$$

It is easy for subunit to obtain feasible solution groups, Y_{1optim}^1, Y_{2optim}^1, using a "flat" string-based evolutionary design process.

Step 2:

(a) The Y^1_{1optim}, Y^1_{2optim} values are grouped according to their cooperative

parameter, e.g., the lubricant of an engine.

(b) The system returns to the granularity Layer 0, and its new feasible

solutions, X^{00}_{optim}, are obtained based on the model described by

Equations (6-2) and (6-3), in which Y^1_{1optim}, Y^1_{2optim} are fixed as the

average value of each group according to the population-based
parameter cooperative mechanism. Attention, in this step, the current
objective function and design constraints adopted in layer 0 are both
transformed to those in the granularity layer 1.

Step 3:

Returning to Step 1(b) to obtain Y^{11}_{1optim}, Y^{11}_{2optim}, with the single sample,

X^0_{optim}, replaced by the sample population, X^{00}_{optim}, the Steps 1 and 2 are

kept performing under the population-based parameter cooperative
mechanism until feasible solutions for both the entire set and the local
subsystems are achieved.

(4) If the design model is developed to include a granularity layer 2, then
the design samples for the newly emerged units are

$$\left\{ \begin{matrix} Z_{11} \\ \vdots \\ Z_{1n} \end{matrix} \right\}^2, \left\{ \begin{matrix} Z_{21} \\ \vdots \\ Z_{2n} \end{matrix} \right\}^2 \text{, which are denoted as } Z_1^2, Z_2^2.$$

The objective functions are

$$\begin{cases} f_{12}(X^0, Y_1^1, Z_1^2) = W^{12} \\ f_{22}(X^0, Y_1^1, Z_2^2) = W^{22} \\ \vdots \end{cases}$$

and $(W^{12}+W^{22}+\ldots\ldots+W^{n2})=W^2$ (6-4)

This is more precise than the mathematical model in granularity layer 1.

Besides the existing constraints, newly emerged constraints are

$$\begin{cases} g_{12}(Y_1^1, X^0, Z_1^2) = 0 \\ h_{12}(Y_1^1, X^0, Z_1^2) < 0 \\ g_{22}(Y_2^1, X^0, Z_2^2) = 0 \\ h_{22}(Y_2^1, X^0, Z_2^2) < 0 \\ \ldots \end{cases}$$ (6-5)

and they include the design samples from various layers.

(5) Evolutionary algorithms for the three-layer model involve a five-step iterative procedure.

Step 1:

$X^0{}_{optim}$, Y_{1optim}^1, Y_{2optim}^1 are obtained according to the above item(3), and

then they are fixed to evolve $Z_1^2, Z_2^2 \cdots Z_n^2$ and therefore,

$Z_{1optim}^2, Z_{2optim}^2 \cdots Z_{noptim}^2$ are obtained.

Step 2:

$Z^2_{1optim}, Z^2_{2optim} \cdots Z^2_{noptim}$ and $X^0{}_{optim}$ are encapsulated to obtain Y^{11}_{1optim},

Y^{11}_{2optim}. (The objective function and constraints are modified, and the parameter cooperative mechanism is adopted.)

Step 3:

$Z^2_{1optim}, Z^2_{2optim} \cdots Z^2_{noptim}$, Y^{11}_{1optim}, Y^{11}_{2optim} are encapsulated to obtain

$X^{00}{}_{optim}$. (The objective function and constraints are modified and become more complex, and the parameter cooperative mechanism is adopted.)

Step 4:

$Z^2_{1optim}, Z^2_{2optim} \cdots Z^2_{noptim}$ and $X^{00}{}_{optim}$ are encapsulated to evolve Y^{111}_{1optim},

Y^{111}_{2optim}. (The objective function and constraints are modified, and the parameter cooperative mechanism is adopted.)

Step 5:

Returning to Step 1 we evolve $Z^2_1 \cdots Z^2_n$ and obtain

$Z^{22}_{1optim}, Z^{22}_{2optim} \cdots Z^{22}_{noptim}$. The procedure repeats until feasible solutions for both the entire set and the local subsystems are obtained.

6.6 Hierarchical cooperative evolutionary design for an engine and its results

As mentioned above, the tribological design for the engine considered in this book is aimed at decreasing the frictional power loss while simultaneously satisfying various design constraints. The simulating functions of frictional power loss are obtained by experiments or by empirical and numerical analysis on the frictional pairs. Both the entire set of samples and the local design samples contribute to the system's frictional power loss, however, they play different roles. The collective basic parameters always dominate in the frictional power loss and the local design samples from different subsystems are always non-correlative with each other. Therefore, a hierarchical cooperative gene model is adopted for the entire set of the engine.

Evolutionary design based on a one-layer hierarchical model

When the internal combustion engine is described by a one-layer hierarchical model, *i.e.*, only the granularity layer 0 exists, the design model will be coarse, and involve fewer design parameters and constraints. It is described by the following design parameters.

Design parameters.
    Viscosity grade of lubricant.
    Stroke–bore ratio.
    Ratio of the stroke to the length of the connecting rod
    Normal rotational speed of the engine

Objective functions. The objective functions are the simulating functions of the system frictional power loss based on the collective basic parameters of the engine. They are listed in Appendix A, and are mainly obtained from References [10-12].

Design constraints. Since the design model does not involve any friction pairs, the tribological constraints have no effect on the design parameters. Design constraints in the granularity layer 0 are simple, and may be classified into two groups,

(1) Functional constraints that aim to keep the basic properties of the engine unchanged, *e.g.* keep the power and basic structure of the engine unchanged.

(2) Parameter value field constraints, *i.e.* parameter must be kept in a value range suggested by the domain experts.

Therefore, the gene sample based on a one-layer model is shown in the following table.

Table 6-1. A gene sample in granularity layer 0.

Design parameters	Viscosity grade of lubricant	Stroke–bore ratio	Ratio of stroke to the length of the connection role	Normal rotational speed of the engine
Gene sample (an example)	110	011	100	111

Because the objective function and the design constraints are simple and coarse, they cannot represent many conflicts between the design parameters. Therefore, feasible solutions based on the one-layer model are easy to obtain during the evolutionary process. Some of these are shown in Table 6-2. Their distribution is described by the objective function value, M, and the constraint satisfying index, Y, in the solution space, as shown in Fig.6-2.

Table 6-2   Some feasible solutions based on the one-layer model

Viscosity grade of lubricant	Stroke -bore Ratio	Ratio of the stroke to the length of the connecting rod	Normal rotational speed of the engine (r/min)	Value of the objective function, M
30	1.0	0.30	2900	0.782581
40	1.15	0.24	2800	0.773639
30	1.15	0.24	2900	0.825715
10W/30	1.15	0.26	3100	0.790330
5W/30	1.2	0.29	3400	0.855843
10W	1.2	0.29	3100	0.813495
40	1.35	0.23	2800	0.849964
10W/30	1.05	0.26	3000	0.782581

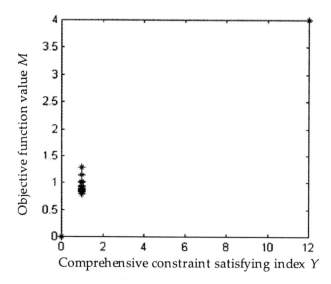

Fig.6-2   Design results in solution space based on granularity layer 0 model

Evolutionary results show that samples with lower viscosity lubricant, lower stroke–bore ratio, lower ratio of the stroke to the length of the connecting rod, and lower normal rotational speed are evolved as the most fit solutions. Among these, the best result is, 10 W, 1.0, 0.24, and

2800 r/min, in which each parameter has been evolved to the minimum limit of its value field, corresponding with the lowest frictional power loss.

It is clear that the best result is meaningless for engineering practice due to the simple design model. Therefore, it is necessary to develop the design model to include a higher granularity layer and describe the design problem with more parameters and constraints.

Cooperating evolutionary design based on a two-layer model

A two-layer hierarchical cooperative evolutionary model is established for the tribological design of the engine featured in this book, which is made up of one granularity layer 0 unit (including the entire sample set) and four granularity layer 1 subunits (*i.e.* the main bearing subsystem, the connecting end bearing subsystem, the piston assembly subsystem, and the valve train subsystem). In each layer, the traditional "flat" gene string model is adopted for the different subsystems and the entire engine set.

The two-layer model for an engine tribological design is described by the followings.

Design parameters.
For the unit in the granularity layer 0
        Viscosity grade of lubricant
        Stroke–bore ratio
        Ratio of stroke to the length of the connecting rod
        Normal rotational speed of the engine

For the piston assembly subsystem in the granularity layer 1

> Viscosity grade of lubricant
>
> Length of skirt
>
> Compression length
>
> Width of rings 1, 2, and 3.
>
> Surface roughness of the cylinder liner
>
> Ellipse rate of the top and bottom of the skirt

For the main bearing subsystem and the big end bearing subsystem in granularity layer 1

> Viscosity grade of lubricant
>
> Width-diameter ratio of the bearing
>
> Width of the bearing
>
> Journal-bushing clearance
>
> Surface roughness of the bush and journal

For the valve train subsystem in granularity layer 1

> Viscosity grade of lubricant
>
> Cam basic circle radius
>
> Parameters P ,Q, R, and S of the polynomial cam
>
> Tolerance of the input guide way
>
> Tolerance of input the guide role
>
> Tolerance of the output guide way
>
> Tolerance of the output guide role

Objective functions. The objective functions are the simulating functions for the system's frictional power loss. Because the design model is becoming increasingly complicated, the simulating functions are based on the lubrication analysis programs. The basic formulas have been listed in Chapter 2 of this book.

Design constraints. The tribological constraints on the engine are multi-disciplinary. They have been classified into three groups, and are summarized in Chapter 5. The detailed describing formulas are also listed in Chapter 2.

As it has been developed to granularity layer 1, the design of the engine has become a multi-parameter, multi-constraint, and multi-disciplinary problem. Its design process has become a hierarchical cooperative evolutionary process. The distribution of the entire set of samples in the solution space during the evolutionary process, is shown in Fig. 6-3. Referring to Fig. 6-2, it can be concluded that, owing to the more precise model and stricter constraints, the number of feasible solutions is reduced, which means that newly expanded subsystems and newly emerged constraints have modified the solution space.

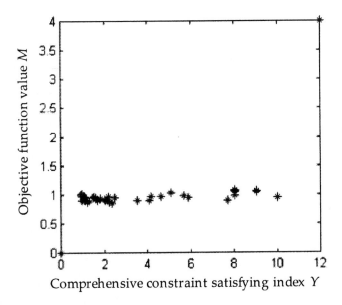

Fig 6-3. The entire set samples in the solution space based on the two-layer model. Newly emerged constraints have modified the solution space compared to that in Fig.6-2.

Table 6-3. The evolved results of the entire set of samples based on the two-layer model

Viscosity grade of lubricant	Stroke –bore ratio	Ratio of the stroke to length of the connecting rod	Normal rotational speed of engine (r/min)	Objective function value, M	Constraint-satisfying index, Y
40	1.00	0.30	3,300	1.000	1.000
30	1.05	0.28	3,300	0.9058	1.000
30	1.05	0.24	3,400	0.96444	1.000
30	1.00	0.29	3,500	0.88954	1.000

The evolved results of the entire set of samples are listed in Table 6-3, in which the first line is the original design scheme for the 4105Q engine.The evolved feasible solutions for the four subsystems are also obtained simultaneously based on the population-based parameter cooperating mechanism, and these are listed in Tables 6-4 to 6-7.

The results show that the evolved feasible solutions are compatible with each other with the same lubricant viscosity grade of 30. The evolved entire set of samples is very close to those of the initial design samples. It can also be concluded that the modifying measures for the initial design scheme of the 4105Q engine included decreasing the viscosity of the lubricant and increasing the stroke–bore ratio. Using these measures, the normal rotational speed of the engine may increases slightly even under the demand of lower frictional power loss.

Furthermore, in the intelligent system for tribological design of engines, these compatible and feasible design samples are recommended to the domain experts for them to examine and decide whether the design solutions are effective in practice. Therefore, the suggested solutions form the foundations for the intelligent system to enable a comprehensive design decision for the entire set of the engine.

Table 6-4. The feasible solutions of the main bearing subsystem.

Design parameters	Viscosity grade of lubricant	Ratio of the width to the diameter of the bearing	Width of the bearing (m)	Clearance (m)	Surface roughness (μm)
Evolved values	30	0.33	0.026	0.0015	0.2

Table 6-5. The feasible solutions of the big end bearing subsystem.

Design parameters	Viscosity grade of lubricant	Ratio of the width to the diameter of the bearing	Width of the bearing (m)	Clearance (m)	Surface roughness (μm)
Evolved values	30	0.424	0.029	0.00125	0.4

Table 6-6. The feasible solutions of the piston assembly subsystem.

Design parameters	Viscosity grade of lubricant	Length of skirt (mm)	Compression length (mm)	Width of ring 1 (mm)	Width of ring 2 (mm)	Width of ring 3 (mm)	Surface roughness (μm)	Ellipse rate of skirt top	Ellipse rate of bottom
Evolved values	30	63.5	74	2	1.9	3.5	0.8	0.34	0.205

Table 6-7. The feasible solutions of the cam and tappet subsystem.

Design parameters	Viscosity grade of lubricant	Basic circle radius of the cam (mm)	Parameter P of the cam	Parameter Q of the cam	Parameter R of the cam	Parameter S of the cam
Evolved value	30	19.9	12	20	36	40

Design parameters	Tolerance of the input guide way (mm)	Tolerance of the input guide role (mm)	Tolerance of the output guide way (mm)	Tolerance of the output guide role (mm)		
Evolved value	0.17	-0.1017	0.17	-0.1018		

6.7 Summary

For the evolutionary design of the entire set of the engine, a long "flat" gene model is not a proper knowledge model. This complicated system requires gene models designed with a complex structure and complex algorithms. In this chapter, the design problem of such a complicated system has been described according the concept of granularity. Then, a hierarchical cooperative evolutionary design method was proposed, which includes three components:
(1) the model evolves along the granularity layers,
(2) the parameters and constraints evolve as the model is developed, and
(3) the design algorithms in each developed node are carried out by evolutionary computation.

A population-based parameter cooperating mechanism was also presented. This coordinates the solutions from the different layers or different subsystems with each other to find compatible, feasible solutions for the entire set of the engine. In this chapter, the application results for the tribological design of an engine were presented. The improved feasible solutions form the foundation for the intelligent system to make comprehensive decision-makings for the entire set of the engine.

References

1. M. A. Rosenman. An exploration into evolutionary models for non-routine design. *Artificial Intelligence in Engineering*. 1997, vol. 11, pp. 287–293
2.. D. A. Coley. An introduction to genetic algorithms for scientists and engineers. World Scientific Publishing, Singapore, 1999
3. J. S. Gero, and T. Schnier, Evolving representation of design cases and

their use in creative design. *Computational methods of creative design.* University of Sydney, 1995

4. M. A. Rosenman and J. G. Gero. Evolving designs by generating useful complex gene structures. *Evolutionary Design by Computers,* London,1999

5. P. Hajela. Non-gradient methods in multidisciplinary design optimization-status and potential. Journal of Aircraft, Vol.36, No. 1, 1999, pp. 255–265

6. D. Dasgupta, and Z. Michalewicz. Evolutionary Algorithms in engineering applications. Springer-Verlag, Berlin, 1997

7. A. Osyczka. Evolutionary Algorithms for single and Multicriteria design optimization. Physica-Verlag, Heidelberg; New York, 2002

8. X. Renbin. The nested modeling support. *Journal of Huazhong University of Science and Technology,* Vol. 24, No. 2, 1996

9. D. R. Adams. Design and analysis: A perspective for the future. Vehicle Tribology, D. Dowson, C. M. Taylor and M. Godet (Eds.), Tribology series, 18, Elsevier, 1991, pp 7–15

10. C. M. Taylor. Automobile engine tribology—design consideration for efficiency and durability. *Wear.* Vol. 22, No. 12, 1994, 40–45

11. E. Ciulli. A review of internal combustion engine losses, Part 1: specific studies on the motion of piston, valves and bearings. *Proceedings of the Institute of Mechanical Engineers, Part D: Journal of Automobile Engineering,* vol. 206, 1992, pp. 223–236

12. E. Ciulli. A review of internal combustion engine losses, Part 2: studies for global evaluations. *Proceedings of the Institute of Mechanical Engineers, Part D: Journal of Automobile Engineering,* vol. 207, 1993, 229–240

Appendix   A

Simulation function between the frictional power loss and the collective basic parameters of engine

The frictional power loss of an engine is the sum of that of the several main friction pairs[1-4].

(1) Frictional power loss of bearing subsystem:

$$N_{fc} = [1.21 \times 10^{-13} \sqrt{SG\eta n^{2.5}} + 5.69 \times 10^{-11} D\sqrt{P_g \eta n^{1.5}}]$$

$$\times [i\sqrt{l_c d_c^3} + \sqrt{d_{cm}^3 \sum l_{mk}}] \qquad \text{(KW)} \qquad \text{(A-1)}$$

where,

    G — equivalent mass of piston and connecting rod(kg);

$$G \approx G_p \sqrt{1 + 0.66\frac{G_c}{G_p} + 1.22(\frac{G_c}{G_p})^2}$$

    G_p—— mass of piston assembly (kg)

    G_c—— mass of the connecting rod group (kg)

    d_c—— diameter of connecting rod bearing (mm)

    l_c—— length of connecting rod bearing (mm)

    d_{cm}—— diameter of main bearing (mm)

l_{cm} —— length of main bearing (mm)

D —— diameter of cylinder liner (mm)

S —— stroke of piston (mm)

i —— cylinder number of engine

P_g —— average cylinder pressure (Mpa);

n —— normal rotational speed of engine (r/min);

η —— dynamic viscosity of lubricant (Pa.s)

(2) Fictional power loss of piston ring and cylinder liner:

$$N_{fr} = 1.03 \times 10^{-7} iDSh(P_g + K_r P_d) fn \quad \text{(KW)} \qquad \text{(A-2)}$$

where, h —— height of piston ring (mm)

K_r —— number of compression rings

P_d —— spring force of piston ring (Mpa)

f —— friction coefficient between ring and cylinder liner, f=0.1~0.15

(3) Frictional power loss due to the inertia mass of piston:

$$N_{fpj} = 2.41 \times 10^{-13} iS^2 \sqrt{DG_A \lambda \eta n^{2.5}} \quad \text{(KW)} \qquad \text{(A-3)}$$

where,

G_A —— reciprocating motion mass of connecting rod (kg)

λ —— ratio of the stroke to connecting rod

(4) Frictional power loss due to the gas pressure:

$$N_{fpq} = 1.27 \times 10^{-11} i \sqrt{P_g D^3 S^3 \lambda \eta n^{1.5}} \quad (KW) \qquad (A\text{-}4)$$

(5) Frictional pressure loss of the valve train subsystem

$$P_{meq} = 2.44 \times 10^{-4} \frac{n_b n}{D^2 Si} + 4120 + \frac{133}{600} \frac{(1 + 1000/n)n_v}{Si}$$

$$+ \frac{0.005}{0.0227} \frac{n_v n}{Si} + \frac{6.32}{15.8} \frac{l_v^{1.5} n_v n^{0.5}}{DSi} + \frac{10700}{42800} \frac{(1 + 1000/n)L_v n_v}{Si} \quad (Pa)$$

$$(A\text{-}5)$$

where, n_b —— bearing numbers;

n_v —— valve numbers;

L_v —— lift range of vale

The corresponding frictional power loss is,

$$N_v = \frac{P_{meq} \times V_h \times i \times n}{300\tau} \qquad (KW) \qquad (A\text{-}6)$$

where, τ —— stroke number.

V_h —— the swept volume of engine

References

1. C.M. Taylor. Automobile engine tribology – design consideration for efficiency and durability. *Wear*. Vol.22, No.12, 1994 : 40-45

2. E. Ciulli. A review of internal combustion engine losses, Part1: specific studies on the motion of piston, valves and bearings. *Proceedings of the Institute of mechanical engineers, Part D: Journal of automobile Engineering.*Vol.206, 1992: 223-236

3. E. Ciulli. A review of internal combustion engine losses, Part2: studies for global evaluations. *Proceedings of the Institute of mechanical engineers, Part D: Journal of automobile Engineering.*Vol.207, 1993: 229-240

4. K. Hamai. Development of a friction predication model for high performance engines. *Lubrication engineering.* July, 1991: 567-573

Subject Index